Materials Science and Engineering

Materials Science and Engineering

Eden Sparks

NY RESEARCH
P R E S S

New York

Published by NY Research Press
118-35 Queens Blvd., Suite 400,
Forest Hills, NY 11375, USA
www.nyresearchpress.com

Materials Science and Engineering
Eden Sparks

International Standard Book Number: 978-1-63238-869-8 (Hardback)

Cataloging-in-Publication Data

Materials science and engineering / Eden Sparks.
  p. cm.
Includes bibliographical references and index.
ISBN 978-1-63238-869-8
1. Materials science. 2. Engineering. 3. Physical sciences.
4. Materials. I. Sparks, Eden.
TA403 .M38 2022
620.11--dc23

Table of Contents

Preface

This book has been written, keeping in view that students want more practical information. Thus, my aim has been to make it as comprehensive as possible for the readers. I would like to extend my thanks to my family and co-workers for their knowledge, support and encouragement all along.

The design and discovery of new materials, specifically solids, is referred to as materials science and engineering. It is an interdisciplinary field that encompasses various fields including physics, chemistry and engineering. It also includes metallurgy, ceramics, solid state physics and chemistry. Material science is concerned with understanding how the structure, performance and properties of a material are affected by its processing history. The understanding of the relationship between various characteristics of the material is known as materials paradigm. The different materials studied within this field are broadly classified into crystalline and amorphous materials. Some of the major types of materials studies within this field are nanomaterials, biomaterials, semiconductors, metals and ceramics. This textbook attempts to understand the multiple branches that fall under the discipline of materials science and engineering, and how such concepts have practical applications. It aims to shed light on some of the unexplored aspects of this field. This book will provide comprehensive knowledge to the readers.

A brief description of the chapters is provided below for further understanding:

Chapter – Introduction

Materials science is a scientific branch which is involved in the design and discovery of new materials. It generally encompasses elements of physics, chemistry and engineering. The topics elaborated in this chapter will help in gaining a better perspective about various aspects of materials science.

Chapter – Sub-disciplines of Materials Science

Materials science can be divided into various sub-disciplines. Some of them are ceramic engineering, crystallography, materials informatics, polymer engineering, polymer science and metallurgy. The chapter closely examines these sub-disciplines of materials science to provide an extensive understanding of the subject.

Chapter – Semiconductors: Principles and Concepts

A material whose electrical conductivity value lies between that of a conductor and an insulator is referred to as a semiconductor. The major types of semiconductors are intrinsic semiconductors and extrinsic semiconductors. This chapter has been carefully written to provide an easy understanding of these types of semiconductors.

Chapter – Polymers

Polymer is a large molecule made up of various repeated subunits. The important areas of study related to polymers are polymer synthesis, polymer characterization, polymer degradation and copolymers. The topics elaborated in this chapter will help in gaining a better perspective about these aspects of polymers.

Chapter – Ceramics and Composites

A ceramic is an inorganic and non-metallic solid composed of metal, ionic, covalent or non-metal bonds. A composite material is a material that is made up of two or more constituent materials with different physical or chemical properties. The chapter closely examines the key concepts of ceramics and composites to provide an extensive understanding of the subject.

Chapter – Nanomaterials and its Types

Nanomaterials are the materials whose size lies between 1 and 1000 nanometers. A few of the main types of nanomaterials are nanoparticles, carbon nanotube, graphene, nanocomposites and nanostructured films. This chapter has been carefully written to provide an easy understanding of these types of nanomaterials.

Eden Sparks

1

Introduction

Materials science is a scientific branch which is involved in the design and discovery of new materials. It generally encompasses elements of physics, chemistry and engineering. The topics elaborated in this chapter will help in gaining a better perspective about various aspects of materials science.

MATERIALS SCIENCE

Materials science is the study of the properties of solid materials and how those properties are determined by a material's composition and structure. It grew out of an amalgam of solid-state physics, metallurgy, and chemistry, since the rich variety of materials properties cannot be understood within the context of any single classical discipline. With a basic understanding of the origins of properties, materials can be selected or designed for an enormous variety of applications, ranging from structural steels to computer microchips. Materials science is therefore important to engineering activities such as electronics, aerospace, telecommunications, information processing, nuclear power, and energy conversion.

The many materials studied and applied in materials science are usually divided into four categories: metals, polymers, semiconductors, and ceramics.

Materials for Energy

An industrially advanced society uses energy and materials in large amounts. Transportation, heating and cooling, industrial processes, communications—in fact, all the physical characteristics of modern life—depend on the flow and transformation of energy and materials through the techno-economic system. These two flows are inseparably intertwined and form the lifeblood of industrial society. The relationship of materials science to energy usage is pervasive and complex. At every stage of energy production, distribution, conversion, and utilization, materials play an essential role, and often special materials properties are needed. Remarkable growth in the understanding of the properties and structures of materials enables new materials, as well as improvements of old ones, to be developed on a scientific basis, thereby contributing to greater efficiency and lower costs.

Classification of Energy-related Materials

Energy materials can be classified in a variety of ways. For example, they can be divided into materials that are passive or active. Those in the passive group do not take part in the actual energy-conversion process but act as containers, tools, or structures such as reactor vessels, pipelines, turbine blades, or oil drills. Active materials are those that take part directly in energy conversion—such as solar cells, batteries, catalysts, and superconducting magnets.

Another way of classifying energy materials is by their use in conventional, advanced, and possible future energy systems. In conventional energy systems such as fossil fuels, hydroelectric generation, and nuclear reactors, the materials problems are well understood and are usually associated with structural mechanical properties or long-standing chemical effects such as corrosion. Advanced energy systems are in the development stage and are in actual use in limited markets. These include oil from shale and tar sands, coal gasification and liquefaction, photovoltaics, geothermal energy, and wind power. Possible future energy systems are not yet commercially deployed to any significant extent and require much more research before they can be used. These include hydrogen fuel and fast-breeder reactors, biomass conversion, and superconducting magnets for storing electricity.

Classifying energy materials as passive or active or in relation to conventional, advanced, or future energy systems is useful because it provides a picture of the nature and degree of urgency of the associated materials requirements. But the most illuminating framework for understanding the relation of energy to materials is in the materials properties that are essential for various energy applications. Because of its breadth and variety, such a framework is best shown by examples. In oil refining, for example, reaction vessels must have certain mechanical and thermal properties, but catalysis is the critical process.

Applications of Energy-related Materials

High-temperature Materials

In order to extract useful work from a fuel, it must first be burned so as to bring some fluid (usually steam) to high temperatures. Thermodynamics indicates that the higher the temperature, the greater the efficiency of the conversion of heat to work; therefore, the development of materials for combustion chambers, pistons, valves, rotors, and turbine blades that can function at ever-higher temperatures is of critical importance. The first steam engines had an efficiency of less than 1 percent, while modern steam turbines achieve efficiencies of 35 percent or more. Part of this improvement has come from improved design and metalworking accuracy, but a large portion is the result of using improved high-temperature materials. The early engines were made of cast iron and then ordinary steels. Later, high-temperature alloys containing nickel, molybdenum, chromium, and silicon were developed that did not melt or fail at temperatures above 540° C (1,000° F). But modern combustion processes are nearing the useful temperature limits that can be achieved with metals, and so new materials that can function at higher temperatures—particularly intermetallic compounds and ceramics—are being developed.

The structural features that limit the use of metals at high temperatures are both atomic and electronic. All materials contain dislocations. The simplest of these are the result of planes of atoms that do not extend all through the crystal, so that there is a line where the plane ends that has fewer atoms than normal. In metals, the outer electrons are free to move. This gives a delocalized cohesion so that, when a stress is applied, dislocations can move to relieve the stress. The result is that metals are ductile: not only can they be easily worked into desired shapes, but when stressed they will gradually yield plastically rather than breaking immediately. This is a desirable feature, but the higher the temperature, the greater the plastic flow under stress—and, if the temperature is too high, the material will become useless. In order to get around this, materials are being studied in which the motion of dislocations is inhibited. Ceramics such as silicon nitride or silicon carbide and intermetallics such as nickel aluminide hold promise because the electrons that hold them together are highly localized in the form of valence or ionic bonds. It is as if metals were held together by a slippery glue while in nonmetals the atoms were connected by rigid rods. Dislocations thus find it much harder to move in nonmetals; raising the temperature does not increase dislocation motion, and the stress needed to make them yield is much higher. Furthermore, their melting points are significantly higher than those of metals, and they are much more resistant to chemical attack. But these desirable features come at a price. The very structure that makes them attractive also makes them brittle; that is, they do not flow when subject to a high stress and are prone to failure by cracking. Modern research is aimed at overcoming this lack of ductility by modification of the material and how it is made. Hot pressing of ceramic powders, for example, minimizes the number of defects at which cracks can start, and the addition of small amounts of certain metals to intermetallics strengthens the cohesion among crystal grains at which fractures normally develop. Such advances, along with intelligent design, hold the promise of being able to build heat engines of much higher efficiency than those now available.

Diamond Drills

Diamond drill bits are an excellent example of how an old material can be improved. Diamond is the hardest known substance and would make an excellent drill bit except that it is expensive and has weak planes in its crystal structure. Because natural diamonds are single crystals, the planes extend throughout the material, and they cleave easily. Such cleavage planes allow a diamond cutter to produce beautiful gems, but they are a disaster for drilling through rock. This limitation was overcome by Stratapax, a sintered diamond material developed by the General Electric Company of the United States. This consists of synthetic diamond powder that is formed into a thin plate and bonded to tungsten-carbide studs by sintering (fusing by heating the material below the melting point). Because the diamond plate is polycrystalline, cleavage cannot propagate through the material. The result is a very hard bit that does not fail by cleavage when it is used to drill through rock to get at oil and natural gas.

Oil Platforms

An important example of dealing with old problems by modern methods is provided by the prevention of crack growth in offshore oil-drilling platforms. The primary structure consists of welded steel tubing that is subject to continually varying stress from ocean waves. Since the cost of building and deploying a platform can amount to several billion dollars, it is imperative that the platform have a long life and not be lost because of premature metal failure.

In the North Sea, 75 percent of the waves are higher than two metres (six feet) and exert considerable stresses on the platform. Cyclic loading of a metal ultimately results in fatigue failure in which surface cracks form, grow over time, and eventually cause the metal to break. Welds are the weak spots for such a process because weld metal has mechanical properties that are inferior to steel, and these are made even worse by internal stresses and defects (such as tiny voids and oxide particles) that are introduced in the welding process. Furthermore, the tube geometry at the weld consists of T- and K-shaped joints, which are natural stress concentrators. Fatigue failure in oil platforms therefore takes place at welds.

Fatigue occurs because cyclic stress causes dislocations to form and to move back and forth in the metal. Dislocation motion can be impeded by the presence of barriers such as small voids, grain boundaries, other dislocations, impurities, or even the surface itself. When dislocations are thereby pinned down, they stop the motion of other dislocations created by the stress, and a tangled dislocation network forms that results in a hard spot in the weld. The stress is then not easily relieved, and types of dislocation motion that are characteristic of the fatigue process initiate a crack at the weld surface. This phenomenon is a direct result of the microstructure of the weld and could be minimized by making the weld very uniform, preferably of the same material as the tubing, and having a very gently curved geometry at the joint. But, in spite of the sophistication of modern welding techniques, this is not yet feasible. An alternate strategy is therefore used in which the progress of the weld crack is monitored so that repairs can be made in time to avoid catastrophic failure. This can be done because, given the geometry of the joint, the depth of the crack is proportional to time until the crack is quite large. By contrast, in laboratory tests in which simple strips of metal are subject to cyclic stress, the growth rate increases as the crack becomes larger. In the T or K configuration in oil platforms, stress is much more evenly distributed, and the crack does not grow at an increasing speed until it is close to being fatal.

A technique for measuring the crack depth is based on the skin effect, the phenomenon in which a high-frequency alternating current is confined to the surface of a conductor. This makes it possible to measure the surface area of a small region with a simple meter, since an increase in crack depth means an increase in current path, and this in turn causes an increase in voltage drop. Measurement over time then allows the time to failure to be estimated; repairs can be effected before failure occurs. In this case, a knowledge of microstructure, the materials science of fatigue, and the study of crack formation have led to a simple testing technique of great economic importance.

Mathematical modeling of mass motion and heat transfer (including convection), along with studies of solidification, gas dissolution, and the effects of fluxes, are providing a much more detailed understanding of the factors controlling weld structure. With this knowledge, it should be possible to make welds with far fewer defects.

Radioactive Waste

A different example is provided by the disposal of radioactive waste. Here the issue is primarily safety and the perception of safety rather than economics. Waste disposal will continue to be one of the factors that inhibit the exploitation of nuclear power until the public perceives it as posing no danger. The current plan is to interpose three barriers between the waste and human beings by first encapsulating it in a solid material, putting that in a metal container, and finally burying that container in geologically stable formations. The first step requires an inert, stable material that will hold the radioactive atoms trapped for a very long time, while the second step requires a material that is highly resistant to corrosion and degradation.

There are two good candidates for encapsulation. The first is borosilicate glass; this can be melted with the radioactive material, which then becomes a part of the glass structure. Glass has a very low solubility, and atoms in it have a very low rate of migration, so that it provides an excellent barrier to the escape of radioactivity. However, glass devitrifies at the high temperatures resulting from the heat of radioactive decay; that is to say, the amorphous glassy state becomes crystalline, and, during this process, many cracks form in the material so that it no longer provides a good barrier against the escape of radioactive atoms. (This problem is more severe in rock than in salt formations, because salt has higher thermal conductivity than rock and dissipates the heat more easily). The problem can be eased by storing the waste above ground for a decade or so. This would allow the initially high rate of decay to decrease, thereby lowering the temperature that would be reached after encapsulation. Handled in this way, borosilicate glass would be an excellent encapsulation material for reactor waste that had been aged for a decade or so.

The other candidate is a synthetic rock made of mineral mixtures such as zirconolite and perovskite. These are very insoluble and, in their natural state, are known to have sequestered radioactive elements for hundreds of millions of years. They are crystalline, ceramic materials whose crystal structures allow radioactive atoms to be immobilized within them. They are not subject to devitrification, since they are already crystalline.

Once encapsulated, radioactive waste must be put into canisters that are corrosion-resistant. These can be made of nickel-steel alloys, but the best candidate so far is a titanium material containing small amounts of nickel and molybdenum and traces of carbon and iron. Even though they are meant to be buried in as dry an environment as possible, these metals are tested by immersing them in brine. Tests show that seawater at 250° C (480° F) would corrode away less than one micrometre (one-thousandth of a millimetre, or four ten-thousandths of an inch) of the surface of the titanium material (known as Ti code 12) per year. This remarkable performance is primarily the result of a tough, highly resistant oxide skin that forms on titanium when exposed to oxygen. It would take thousands of years for the canisters to be penetrated by corrosion.

In order to estimate the effectiveness of such waste disposal, it must be noted that the waste is highly radioactive and dangerous initially but that the danger decreases with time. Radioactivity decays to such levels that the danger is much less after a few hundred years, extremely low after 500 years, and negligible after 1,000 years. In order to breach the triple-barrier system, groundwater must migrate to the canister, eat it away, and then leach out the radioactive atoms from the encapsulating glass or ceramic. This is a process that most probably would take far longer than a single millennium. A careful application of materials science can make radioactive waste disposal safer than current disposal methods for other toxic wastes.

Photovoltaics

Photovoltaic systems are an attractive alternative to fossil or nuclear fuels for the generation of electricity. Sunlight is free, it does not use up an irreplaceable resource, and its conversion to electricity is nonpolluting. In fact, photovoltaics are now in use where power lines from utility grids are either not possible or do not exist, as in outer space or remote, nonurban locations.

The barrier to widespread use of sunlight to generate electricity is the cost of photovoltaic systems. The application of materials science is essential in efforts to lower the cost to levels that can compete with those for fossil or nuclear fuels.

The conversion of light to electricity depends on the electronic structure of solar cells with two or more layers of semiconductor material that can absorb photons, the primary energy packets of light. The photons raise the energy level of the electrons in the semiconductor, exciting some to jump from the lower-energy valence band to the higher-energy conduction band. The electrons in the conduction band and the holes they have left behind in the valence band are both mobile and can be induced to move by a voltage. The electron motion, and the movement of holes in the opposite direction, constitute an electric current. The force that drives electrons and holes through a circuit is created by the junction of two dissimilar semiconducting materials, one of which has a tendency to give up electrons and acquire holes (thereby becoming the positive, or p-type, charge carrier) while the other accepts electrons (becoming the negative, or n-type, carrier). The electronic structure that permits this is the band gap; it is equivalent to the energy required to move an electron from the lower band to the higher. The magnitude of this gap is important. Only photons with energy greater than that of the band gap can excite electrons from the valence band to the conduction band; therefore, the smaller the gap, the more efficiently light will be converted to electricity—since there is a greater range of light frequencies with sufficiently high energies. On the other hand, the gap cannot be too small, because the electrons and holes then find it easy to recombine, and a sizable current cannot be maintained.

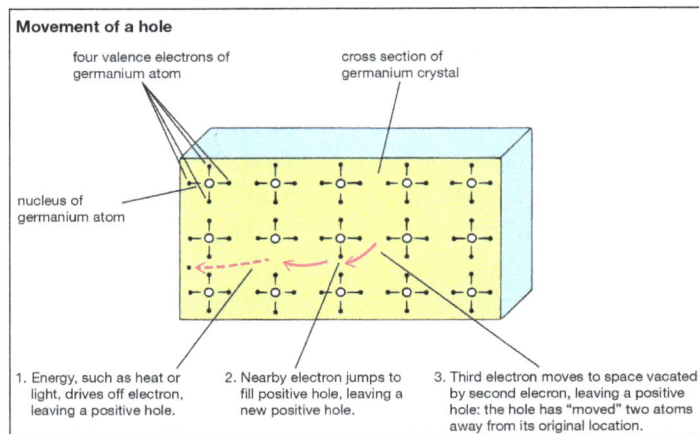

Electron Hole: Movement
Movement of an electron hole in a crystal lattice.

The band gap defines the theoretical maximum efficiency of a solar cell, but this cannot be attained because of other materials factors. For each material there is an intrinsic rate of recombination of electrons and holes that removes their contribution to electric current. This recombination is enhanced by surfaces, interfaces, and crystal defects such as grain boundaries, dislocations, and impurities. Also, a fraction of the light is reflected by the cell's surface rather than being absorbed, and some can pass through the cell without exciting electrons to the conduction band.

Improvements in the trade-off between cell efficiency and cost are well illustrated by the preparation of silicon that is the basic material of current solar cells. Initially, high-purity silicon was grown from a silicon melt by slowly pulling out a seed crystal that grew by the accretion and slow solidification of the molten material. Known as the Czochralski process, this resulted in a high-purity, single-crystal ingot that was then sliced into wafers about 1 millimetre (0.04 inch) thick. Each wafer's surface was then "doped" with impurities to create p-type and n-type materials with a junction between them. Metal was then deposited to provide electrical leads, and the wafer was encapsulated to yield a cell about 100 millimetres in diameter. This was an expensive and time-consuming process; it has been much improved in a variety of ways. For example, high-purity silicon can be made at drastically reduced cost by chemically converting ordinary silicon to silane or trichlorosilane and then reducing it back to silicon. This silane process is capable of continuous operation at a high production rate and with low energy input. In order to avoid the cost and waste associated with sawing silicon into wafers, methods of directly drawing molten silicon into thin sheets or ribbons have been developed; these can produce crystalline, polycrystalline, or amorphous material. Another alternative is the manufacture of thin films on ceramic substrates—a process that uses much less silicon than other methods. Single-crystal silicon has a higher efficiency than other forms, but it is also much more expensive. The materials challenge is to find a combination of cost and efficiency that makes photovoltaic electricity economically possible.

Surface treatments that increase efficiency include deposition of antireflecting coatings, such as silicon nitride, on the front of the cell and highly reflective coatings on the rear. Thus, more of the light that strikes a cell actually enters it, and light that escapes out the back is reflected back into the cell. An ingenious surface treatment is part of the point contact method, in which the surface of the cell is not planar but microgrooved so that light is randomly reflected as it strikes the cell. This increases the amount of light that can be captured by the cell.

Materials for Ground Transportation

The global effort to improve the efficiency of ground transportation vehicles, such as automobiles, buses, trucks, and trains, and thereby reduce the massive amounts of pollutants they emit, provides an excellent contextwithin which to illustrate how materials science functions to develop new or better materials in response to critical human needs. For the automobile industry in particular, the story is a fascinating one in which the desire for lower vehicle weight, reduced emissions, and improved fuel economy has led to intense competition among aluminum, plastics, and steel companies for shares in the enormous markets involved (40 million to 50 million cars and trucks per year worldwide). In this battle, materials scientists have a key role to play because the success of their efforts to develop improved materials will determine the shape and viability of future automobiles.

Just how seriously suppliers to the industry view the need either to protect or to increase their share of these enormous markets is demonstrated by their establishing of special programs, consortia, or centres that are specifically designed to develop better alloys, plastics, or ceramics for automotive applications. For example, in the United States a program at the Aluminum Company of America (Alcoa) called the aluminum intensive vehicle (AIV), and a similar one at Reynolds Metals, were established to develop materials and processes for making automobile "space frames" consisting of aluminum-alloy rods and die-cast connectors joined by welding and adhesive bonding. Not to be outdone, another aluminum company, Alcan Aluminium Limited of Canada, in a program entitled aluminum structured vehicle technology (ASVT), began to investigate the construction of automobile unibodies from adhesively bonded aluminum sheet. The plastics industry, of course, has a powerful interest in replacing as many metal automobile components as possible, and in order to help bring this about a centre called D&S Plastics International was formed in the Detroit, Mich., area of the United States by three corporations. The specific aim of this centre was to develop materials and a process suitable for forming several connected panels or components (*e.g.*, body panels and bumper fascias) simultaneously out of different types of plastics. The centrepiece of the operation was a 4,000-ton co-injection press that could lead to cost reductions as great as 50 percent and thereby make the use of plastics for automotive applications more attractive.

In programs such as these, and in many more carried out by vendors and within the automobile companies themselves, materials scientists with specialized training in advanced metals, plastics, and ceramics have been leading a revolution in the automotive industry. The following sections describe specific needs that have been identified for improving the performance of automobiles and other ground-transportation vehicles, as well as approaches that materials scientists have taken in response to those needs.

Metals

Aluminum

Since aluminum has about one-third the density of steel, its substitution for steel in automobiles would seem to be a sensible approach to reducing weight and thereby increasing fuel economy and reducing harmful emissions. Such substitutions cannot be made, however, without due consideration of significant differences in other properties of the two materials. This is one important facet of the materials scientist's job—to help evaluate the suitability of a material for a given application based on how its properties balance against load and performance requirements specified by the design engineer. In this case (aluminum versus steel), it is instructive to consider the materials scientist's approach to evaluating the use of aluminum in automotive panels—such components as doors, hoods, trunk decks, and roofs that can make up more than 60 percent of a vehicle's weight.

Two primary properties of any metal are (1) its yield strength, defined as its ability to resist permanent deformation (such as a fender dent), and (2) its elastic modulus, defined as its ability to resist elastic or springy deflection like a drum head. By alloying, aluminum can be made to have a yield strength equal to a moderately strong steel and therefore to exhibit similar resistance to denting in an automobile panel. On the other hand, alloying does not normally affect the elastic modulus of metals significantly, so that automotive door panels or hoods made from aluminum alloys, all of which have approximately one-third the modulus of steel, would be floppy and suffer large deflections when buffeted by the wind, for example. From this point of view, aluminum would appear to be a marginal choice for body panels.

One might attempt to overcome this deficiency by increasing the thickness of the aluminum sheet stock to three times the thickness of the steel it is intended to replace. This, however, would simply increase the weight to roughly that of an equivalent steel structure and thus defeat the purpose of the exercise. Fortunately, as was elegantly demonstrated in 1980 by two British materials scientists, Michael Ashby and David Jones, when proper account is taken of the way an actual door panel deflects, constrained as it is by the door edges, it is possible to use aluminum sheet only slightly thicker than the steel it would replace and still achieve equivalent performance. The net result would be a weight savings of almost two-thirds by the substitution of aluminum for steel on such body components. This suggests that understanding the interrelationship between materials properties and structural design is an important factor in the successful application of materials science.

Another important activity of the materials scientist is that of alloydevelopment, which in some cases involves designing alloys for very specific applications. For example, in Alcoa's AIV effort, materials scientists and engineers developed a special casting alloy for use as cast aluminum nodes (connecters) in their space frame design. Ordinarily, metal castings exhibit very little toughness, or ductility, and they are therefore prone to brittle fracture followed by catastrophic failure. Since the integrity of an automobile would be limited by having relatively brittle body components, a proprietary casting alloy and processing procedure were developed that provide a material of much greater ductility than is normally available in a casting alloy.

Many other advances in aluminum technology, brought about by materials scientists and design engineers, have led to a greater acceptance of aluminum in automobiles, trucks, buses, and even light rail vehicles. Among these are alloys for air-conditioner components that are designed to be chemically compatible with environmentally safer refrigerants and to withstand the higher pressures required by them. Also, alloys have been developed that combine good formability and corrosion resistance with the ability to achieve maximum strength without heat treating; these alloys develop their strength during the forming operation. As a consequence, the list of vehicles that contain significant quantities of aluminum substituted for steel has steadily grown. A milestone was reached in 1992 with a limited-edition Jaguar sports car that was virtually all aluminum, including the engine, adhesively bonded chassis, and skin. Somewhat less expensive and in full production were Honda's Acura NSX, containing more than 400 kilograms (900 pounds) of aluminum compared with about 70 kilograms for the average automobile, and General Motors' Saturn, with an aluminum engine block and cylinder heads. These vehicles and others took their place alongside the British Land Rover, which was built with all-aluminum body panels beginning in 1948—a choice dictated by a shortage of steel during World War II and continued by the manufacturer ever since.

Steel

While the goal of the aluminum and plastics industries is to achieve vehicle weight reductions by substituting their products for steel components, the goal of the steel industry is to counter such inroads with such innovative developments as high-strength, but inexpensive, "microalloyed" steels that achieve weight savings by thickness reductions. In addition, alloys have been developed that can be tempered (strengthened) in paint-baking ovens rather than in separate and expensive heat-treatment furnaces normally required for conventional steels.

The microalloyed steels, also known as high-strength low-alloy (HSLA) steels, are intermediate in composition between carbon steels, whose properties are controlled mainly by the amount of carbon they contain (usually less than 1 percent), and alloy steels, which derive their strength, toughness, and corrosion resistance primarily from other elements, including silicon, nickel, and manganese, added in somewhat larger amounts. Developed in the 1960s and resurrected in the late 1970s to satisfy the need for weight savings through greater strength, the HSLA steels tend to be low in carbon with minute additions of titanium or vanadium, for example. Offering tensile strengths that can be triple the value of the carbon steels they are designed to replace (*e.g.,* 700 megapascals versus 200 megapascals), they have led to significant weight savings through thickness reductions—albeit at a slight loss of structural stiffness, because their elastic moduli are the same as other steels. They are considered to be quite competitive with aluminum substitutes for two reasons: they are relatively inexpensive (steel sells for one-half the price of aluminum on a per-unit-weight basis); and very little change in fabrication and processing procedures is needed in switching from carbon steel to HSLA steel, whereas major changes are usually required in switching to aluminum.

Bake-hardenable steels were developed specifically for the purpose of eliminating an expensive fabrication step—*i.e.,* the heat-treating furnace, where steels are imparted with their final strength. To do this, materials scientists have designed steels that can be strengthened in the same ovens used to bake body paint onto the part. These furnaces must operate at relatively low temperatures (170 °C, or 340 °F), so that special steels had to be developed that would achieve suitable strengths at heat-treatment temperatures very much below those normally employed (up to 600 °C, or 1,100 °F).

Knowing that high-alloy steels would never be hardenable at such low temperatures, materials scientists focused their attention on carbon steels, but even here adequate strengths could not be obtained initially. Then in the 1980s scientists at the Japanese Sumitomo Metal Industriesdeveloped a steel containing nitrogen (a gas that constitutes three-quarters of the Earth's atmosphere) in addition to carbon and several other additives. Very high strengths (over 900 megapascals) and excellent toughness can be achieved on formed parts with this inexpensive addition after baking for 20 minutes at temperatures typical for a paint-baking operation.

Plastics and Composites

The motive for replacing the metal components of cars, trucks, and trains with plastics is the expectation of large weight savings due to the large differences in density involved: plastics are one-sixth the weight of steel and one-half that of aluminum per unit volume. However, as in evaluating the suitability of replacing steel with aluminum, the materials scientist must compare other properties of the materials in order to determine whether the tradeoffs are reasonable. For two reasons, the likely conclusion would be that plastics simply are not suitable for this type of application: the strength of most plastics, such as epoxies and polyesters, is roughly one-fifth that of steel or aluminum; and their elastic modulus is one-sixtieth that of steel and one-twentieth that of aluminum. On this basis, plastics do not appear to be suitable for structural components. What, then, accounts for the successful use that has been made of them? The answer lies in efforts made over the years by materials scientists, polymer chemists, mechanical engineers, and production managers to combine relatively weak and low-stiffness resins with high-strength, high-modulus reinforcements, thereby making new materials called composites with much more suitable properties than plastics alone.

The reinforcements used in composites are generally chosen for their high strength and modulus, as might be expected, but economic considerations often force compromises. For example, carbon fibres have extremely high modulus values (up to five times that of steel) and therefore make excellent reinforcements. However, their cost precludes their extensive use in automobiles, trucks, and trains, although they are used regularly in the aerospace industry. More suitable for non-aerospace applications are glass fibres (whose modulus can approach 1.5 times that of aluminum) or, in somewhat special cases, a mixture of glass and carbon fibres.

The physical form and shape of the reinforcements vary greatly, depending on many factors. The most effective reinforcements are long fibres, which are employed either in the form of a woven cloth or as separate layers of unidirectional fibres stacked upon one another until the proper laminate thickness is achieved. The resin may be applied to the fibres or cloth before laying up, thus forming what are termed prepregs, or it may be added later by "wetting out" the fibres. In either case, the assembly is then cured, usually under pressure, to form the composite. This type of composite takes full advantage of the properties of the fibres and is therefore capable of yielding strong, stiff panels. Unfortunately, the labour involved in the lay-up operations and other factors make it very expensive, so that long-fibre reinforcement is used only sparingly in the automobile industry.

One attempt to avoid expensive hand lay-up operations involves chopped fibres that are employed in mat form, somewhat like felt, or as loose fibres that may be either blown into a mold or injected into a mold along with the resin. Another method does not use fibres at all; instead the reinforcement is in the form of small, high-modulus particles. These are the least expensive of all to process, since the particles are simply mixed into the resin, and the mixture is used in various types of molds. On the other hand, particles are the least efficient reinforcement material; as a consequence, property improvements are not outstanding.

In choosing the other major constituent in composites, the polymer matrix, one faces a somewhat daunting variety, including epoxies, polyimides, polyurethanes, and polyesters. Each has its advantages and disadvantages that must be evaluated in order to determine suitability for a particular application. Among the factors to be considered are cost, processing temperature (curing temperature if using a thermoset polymer and melting temperature if using a thermoplastic), flow properties in the molding operation, sag resistance during paint bake out, moisture resistance, and shelf life. The number of combinations of resins, reinforcements, production methods, and fibre-to-resin ratios is so challenging that materials scientists must join forces with polymer chemists and engineers from the design, production, and quality-control departments of the company in order to choose the right combination for the application.

Judging by the inroads that have been made in replacing metals with composites, it appears that technologists have been making the right choices. The introduction of fibreglass-reinforced plastic skins on General Motors' 1953 Corvette sports car marked the first appearance of composites in a production model, and composites have continued to appear in automotive components ever since. In 1984, General Motors' Fiero was placed on the market with the entire body made from composites, and the Camaro/Firebird models followed with doors, roof panels, fenders, and other parts made of composites. Composites were also chosen for exterior panels in the Saturn, which appeared in 1990. In addition, they have had less visible applications—for example, the glass-reinforced nylon air-intake manifold on some BMW models.

Ceramics

Ceramics play an important role in engine efficiency and pollution abatement in automobiles and trucks. For example, one type of ceramic, cordierite (a magnesium aluminosilicate), is used as a substrate and support for catalysts in catalytic converters. It was chosen for this purpose because, along with many ceramics, it is lightweight, can operate at very high temperatures without melting, and conducts heat poorly (helping to retain exhaust heat for improved catalytic efficiency). In a novel application of ceramics, a cylinder wall was made of transparent sapphire (aluminum oxide) by General Motors' researchers in order to examine visually the internal workings of a gasoline engine combustion chamber. The intention was to arrive at improved understanding of combustion control, leading to greater efficiency of internal-combustion engines.

Another application of ceramics to automotive needs is a ceramic sensor that is used to measure the oxygen content of exhaust gases. The ceramic, usually zirconium oxide to which a small amount of yttrium has been added, has the property of producing a voltage whose magnitude depends on the partial pressure of oxygen surrounding the material. The electrical signal obtained from such a sensor is then used to control the fuel-to-air ratio in the engine in order to obtain the most efficient operation.

Because of their brittleness, ceramics have not been used as load-bearing components in ground-transportation vehicles to any great extent. The problem remains a challenge to be solved by materials scientists of the future.

Materials for Aerospace

The primary goal in the selection of materials for aerospace structures is the enhancement of fuel efficiency to increase the distance traveled and the payload delivered. This goal can be attained by developments on two fronts: increased engine efficiency through higher operating temperatures and reduced structural weight. In order to meet these needs, materials scientists look to materials in two broad areas—metal alloys and advanced composite materials. A key factor contributing to the advancement of these new materials is the growing ability to tailor materials to achieve specific properties.

Metals

Many of the advanced metals currently in use in aircraft were designed specifically for applications in gas-turbine engines, the components of which are exposed to high temperatures, corrosive gases, vibration, and high mechanical loads. During the period of early jet engines (from about 1940 to 1970), design requirements were met by the development of new alloys alone. But the more severe requirements of advanced propulsion systems have driven the development of novel alloys that can withstand temperatures greater than 1,000° C (1,800° F), and the structural performance of such alloys has been improved by developments in the processes of melting and solidification.

Melting and Solidifying

Alloys are substances composed of two or more metals or of a metal and a nonmetal that are intimately united, usually by dissolving in each other when they are melted. The principal objectives of melting are to remove impurities and to mix the alloying ingredients homogeneously in the base metal. Major advances have been made with the development of new processes based on melting under vacuum (hot isostatic pressing), rapid solidification, and directional solidification.

In hot isostatic pressing, prealloyed powders are packed into a thin-walled, collapsible container, which is placed in a high-temperature vacuum to remove adsorbed gas molecules. It is then sealed and put in a press, where it is exposed to very high temperatures and pressures. The mold collapses and welds the powder together in the desired shape.

Molten metals cooled at rates as high as a million degrees per second tend to solidify into a relatively homogeneous microstructure, since there is insufficient time for crystalline grains to nucleate and grow. Such homogeneous materials tend to be stronger than the typical "grainy" metals. Rapid cooling rates can be achieved by "splat" cooling, in which molten droplets are projected onto a cold surface. Rapid heating and solidification can also be achieved by passing high-power laser beams over the material's surface.

Unlike composite materials, grainy metals exhibit properties that are essentially the same in all directions, so they cannot be tailored to match anticipated load paths (*i.e.*, stresses applied in specific directions). However, a technique called directional solidification provides a certain degree of tailorability. In this process the temperature of the mold is precisely controlled to promote the formation of aligned stiff crystals as the molten metal cools. These serve to reinforce the component in the direction of alignment in the same fashion as fibres reinforce composite materials.

Alloying

These advances in processing have been accompanied by the development of new "superalloys." Superalloys are high-strength, often complex alloys that are resistant to high temperatures and severe mechanical stress and that exhibit high surface stability. They are commonly classified into three major categories: nickel-based, cobalt-based, and iron-based. Nickel-based superalloys predominate in the turbine section of jet engines. Although they have little inherent resistance to oxidation at high temperatures, they gain desirable properties through the addition of cobalt, chromium, tungsten, molybdenum, titanium, aluminum, and niobium.

Aluminum-lithium alloys are stiffer and less dense than conventional aluminum alloys. They are also "superplastic," owing to the fine grain size that can now be achieved in processing. Alloys in this group are appropriate for use in engine components exposed to intermediate to high temperatures; they can also be used in wing and body skins.

Titanium alloys, as modified to withstand high temperatures, are seeing increased use in turbine engines. They are also employed in airframes, primarily for military aircraft but to some extent for commercial planes as well.

Polymer-matrix Composites

PMCs are of two broad types, thermosets and thermoplastics. Thermosetsare solidified by irreversible chemical reactions, in which the molecules in the polymer "cross-link," or form connected chains. The most common thermosetting matrix materials for high-performance composites used in the aerospace industry are the epoxies. Thermoplastics, on the other hand, are melted and then solidified, a process that can be repeated numerous times for reprocessing. Although the manufacturing technologies for thermoplastics are generally not as well developed as those for thermosets, thermoplastics offer several advantages. First, they do not have the shelf-life problem associated with thermosets, which require freezer storage to halt the irreversible curing process that begins at room temperature. Second, they are more desirable from an environmental point of view, as they can be recycled. They also exhibit higher fracture toughness and better resistance to solvent attack. Unfortunately, thermoplastics are more expensive, and they generally do not resist heat as well as thermosets; however, strides are being made in developing thermoplastics with higher melting temperatures. Overall, thermoplastics offer a greater choice of processing approaches, so that the process can be determined by the scale and rate of production required and by the size of the component.

A variety of reinforcements can be used with both thermoset and thermoplastic PMCs, including particles, whiskers (very fine single crystals), discontinuous (short) fibres, continuous fibres, and textile preforms (made by braiding, weaving, or knitting fibres together in specified designs). Continuous fibres are more efficient at resisting loads than are short ones, but it is more difficult to fabricate complex shapes from materials containing continuous fibres than from short-fibre or particle-reinforced materials. To aid in processing, most high-performance composites are strengthened with filaments that are bundled into yarns. Each yarn, or tow, contains thousands of filaments, each of which has a diameter of approximately 10 micrometres (0.01 millimetre, or 0.0004 inch).

Depending on the application and on the type of load to be applied to the composite part, the reinforcement can be random, unidirectional (aligned in a single direction), or multidirectional (oriented in two or three dimensions). If the load is uniaxial, the fibres are all aligned in the load direction to gain maximum benefit of their stiffness and strength. However, for multidirectional loading (for example, in aircraft skins), the fibres must be oriented in a variety of directions. This is often accomplished by stacking layers (or lamina) of continuous-fibre systems.

The most common form of material used for the fabrication of composite structures is the preimpregnated tape, or "prepreg." There are two categories of prepreg: tapes, generally 75 millimetres (3 inches) or less in width, intended for fabrication in automated, computer-controlled tape-laying machines; and "broad goods," usually several metres in dimension, intended for hand lay-up and large sheet applications. To make prepregs, fibres are subjected to a surface treatment so that the resin will adhere to them. They are then placed in a resin bath and rolled into tapes or sheets.

To fabricate the composite, the manufacturer "lays up" the prepreg according to the reinforcement needs of the application. This has traditionally been done by hand, with successive layers of a broad-goods laminate stacked over a tool in the shape of the desired part in such a way as to accommodate the anticipated loads. However, efforts are now being directed toward automated fibre-placement methods in order to reduce costs and ensure quality and repeatability. Automated fibre-placement processes fall into two categories, tape laying and filament winding. The tape-laying process involves the use of devices that control the placement of narrow prepreg tapes over tooling with the contours of the desired part and along paths prescribed by the design requirements of the structure. The width of the tape determines the "sharpness" of the turns required to place the fibres in the prescribed direction—i.e., wide tapes are used for gradual turns, while narrow tapes are required for the sharp turns associated with more complex shapes.

Filament winding uses the narrowest prepreg unit available—the yarn, or tow, of impregnated filaments. In this process, the tows are wound in prescribed directions over a rotating mandrel in the shape of the part. Successive layers are added until the required thickness is reached. Although filament winding was initially limited to geodesic paths (i.e.,winding the fibres along the most direct route between two points), the process is now capable of fabricating complex shapes through the use of robots.

For thermosetting polymers, the structure generated by either tape laying or filament winding must undergo a second manipulation in order to solidify the polymer through a curing reaction. This is usually accomplished by heating the completed structure in an autoclave, or oven. Thermoplastic systems offer the advantage of on-line consolidation, so that the high energy and capital costs associated with the curing step can be eliminated. For these systems, prepreg can be locally melted, consolidated, and cooled at the point of contact so that a finished structure is produced. A variety of energy sources are used to concentrate heat at the point of contact, including hot-gas torches, infrared light, and laser beams.

Pultrusion, the only truly continuous process for manufacturing parts from PMCs, is economical but limited to the production of beamlike shapes. On a pultrusion line, fibres and the resin are pushed through a heated die, or shaping tool, at one end, then cooled and pulled out at the other end. This process can be applied to both thermoplastic and thermoset polymers.

Resin transfer molding, or RTM, is a composites processing method that offers a high potential for tailorability but is currently limited to low-viscosity (easily flowing) thermosetting polymers. In RTM, a textile preform—made by braiding, weaving, or knitting fibres together in a specified design—is placed into a mold, which is then closed and injected with a resin. After consolidation, the mold is opened and the part removed. Preforms can be made in a wide variety of architectures, and several can be joined together during the RTM process to form a multi-element preform offering reinforcement in specific areas and load directions.

The similarity of meltable thermoplastic polymers to metals has prompted the extension of techniques used in metalworking. Sheet forming, used since the 19th century by metallurgists, is now applied to the processing of thermoplastic composites. In a typical thermoforming process, the sheet stock, or preform, is heated in an oven. At the forming temperature, the sheet is transferred into a forming system, where it is forced to conform to a tool, with a shape that matches the finished part. After forming, the sheet is cooled under pressure and then removed. Stretch forming, a variation on thermoplastic sheet forming, is specifically designed to take advantage of the extensibility, or ability to be stretched, of thermoplastics reinforced with long, discontinuous fibres. In this process, a straight preconsolidated beam is heated and then stretched over a shaped tool to introduce curvature. The specific advantage of stretch forming is that it provides an automated way to achieve a very high degree of fibre-orientation control in a wide range of part sizes.

Metal-matrix and Ceramic-matrix Composites

The requirement that finished parts be able to operate at temperatures high enough to melt or degrade a polymer matrix creates the need for other types of matrix materials, often metals. Metal matrices offer not only high-temperature resistance but also strength and ductility, or "bendability," which increases toughness. The main problems with metal-matrix composites (MMCs) are that even the lightest metals are heavier than polymers, and they are very complex to process. MMCs can be used in such areas as the skin of a hypersonic aircraft, but on wing edges and in engines temperatures often exceed the melting point of metals. For the latter applications, ceramic-matrix composites (CMCs) are seeing increasing use, although the technology for CMCs is less mature than that for PMCs. Ceramics consist of alumina, silica, zirconia, and other elements refined from fine earth and sand or of synthetic materials, such as silicon nitride or silicon carbide. The desirable properties of ceramics include superior heatresistance and low abrasive and corrosive properties. Their primary drawback is brittleness, which can be reduced by reinforcing with fibres or whiskers. The reinforcement material can be a metal or another ceramic.

Unlike polymers and metals, which can be processed by techniques that involve melting (or softening) followed by solidification, high-temperature ceramics cannot be melted. They are generally produced by some variation of sintering, a technique that renders a combination of materials into a coherent mass by heating to high temperatures without complete melting. If continuous fibres or textile weaves (as opposed to short fibres or whiskers) are involved, sintering is preceded by impregnating the assembly of fibres with a slurry of ceramic particles dispersed in a liquid. A major benefit of using CMCs in aircraft engines is that they allow higher operating temperatures and thus greater combustion efficiency, leading to reduced fuel consumption. An additional benefit is derived from the low density of CMCs, which translates into substantial weight savings.

Other Advanced Composites

Carbon-carbon composites are closely related to CMCs but differ in the methods by which they are produced. Carbon-carbon composites consist of semicrystalline carbon fibres embedded in a matrix of amorphouscarbon. The composite begins as a PMC, with semicrystalline carbon fibres impregnated with a polymeric phenolic resin. The resin-soaked system is heated in an inert atmosphere to pyrolyze, or char, the polymer to a carbon residue. The composite is re-impregnated with polymer, and the pyrolysis is repeated. Continued repetition of this impregnation/pyrolysis process yields a structure with minimal voids. Carbon-carbon composites retain their strength at 2,500 °C (4,500 °F) and are used in the nose cones of reentry vehicles. However, because they are vulnerable to oxidation at such high temperatures, they must be protected by a thin layer of ceramic.

While materials research for aerospace applications has focused largely on mechanical properties such as stiffness and strength, other attributes are important for use in space. Materials are needed with a near-zero coefficient of thermal expansion; in other words, they have to be thermally stable and should not expand and contract when exposed to extreme changes in temperature. A great deal of research is focused on developing such materials for high-speed civilian aircraft, where thermal cycling is a major issue. High-toughness materials and nonflammable resin composite systems are also under investigation to improve the safety of aircraft interiors.

Efforts are also being directed toward the development of "smart," or responsive, materials. Representing another attempt to mimic certain characteristics of living organisms, smart materials, with their built-in sensors and actuators, would react to their external environment by bringing on a desired response. This would be done by linking the mechanical, electrical, and magnetic properties of these materials. For example, piezoelectric materials generate an electrical current when they are bent; conversely, when an electrical current is passed through these materials, they stiffen. This property can be used to suppress vibration: the electrical current generated during vibration could be detected, amplified, and sent back, causing the material to stiffen and stop vibrating.

Materials for Computers and Communications

The basic function of computers and communications systems is to process and transmit information in the form of signals representing data, speech, sound, documents, and visual images. These signals are created, transmitted, and processed as moving electrons or photons, and so the basic materials groups involved are classified as electronic and photonic. In some cases, materials known as optoelectronic bridge these two classes, combining abilities to interact usefully with both electrons and photons.

Among the electronic materials are various crystalline semiconductors; metalized film conductors; dielectric films; solders; ceramics and polymers formed into substrates on which circuits are assembled or printed; and gold or copper wiring and cabling.

Photonic materials include a number of compound semiconductors designed for light emission or detection; elemental dopants that serve as photonic performance-control agents; metal- or diamond-film heat sinks; metalized films for contacts, physical barriers, and bonding; and silica glass, ceramics, and rare earths for optical fibres.

Electronic Materials

Between 1955 and 1990, improvements and innovations in semiconductortechnology increased the performance and decreased the cost of electronic materials and devices by a factor of one million—an achievement unparalleled in the history of any technology. Along with this extraordinary explosion of technology has come an exponentially upward spiral of the capital investment necessary for manufacturing operations. In order to maintain cost-effectiveness and flexibility, radical changes in materials and manufacturing operations will be necessary.

Semiconductor Crystals

Silicon

Bulk semiconductor silicon for the manufacture of integrated circuits(sometimes referred to as electronic-grade silicon) is the purest material ever made commercially in large quantities. One of the most important factors in preparing this material is control of such impurities as boron, phosphorus, and carbon. For the ultimate levels of integrated-circuit design, stray contaminant atoms must constitute less than 0.1 part per trillion of the material.

For fabrication into integrated circuits, bulk semiconductor silicon must be in the form of a single-crystal material with high crystalline perfection and the desired charge-carrier concentration. The size of the silicon ingot, or boule, has been scaled up in recent years, in order to provide wafers of increasing diameter that are demanded by the economics of integrated-circuit manufacturing. Most commonly, a 60-kilogram (130-pound) charge is grown to an ingot with a diameter of 200 millimetres (8 inches), but the semiconductor industry will soon require ingots as large as 300 millimetres. The ingots are then converted into wafers by machining and chemical processes.

III–V Compounds

Although silicon is by far the most commonly used crystal material for integrated circuits, a significant volume of semiconductor devices and circuits employs III–V technology, so named because it is based on crystalline compounds formed by combining metallic elements from column III and nonmetallic elements from column V of the periodic table of chemical elements. When the elements are gallium and arsenic, the semiconductor is called gallium arsenide, or GaAs. However, other elements such as indium, phosphorus, and aluminum are often used in the compound to achieve specific performance characteristics.

For electronic applications, the III–V semiconductors offer the basic advantage of higher electron mobility, which translates into higher operating speeds. In addition, devices made with III–V compounds provide lower voltage operation for specific functions, radiation hardness (especially important for satellites and space vehicles), and semi-insulating substrates (avoiding the presence of parasitic capacitance in switching devices).

III–V materials are more difficult to handle than silicon, and a III–V wafer or substrate usually is less than half the size of a silicon wafer. In addition, a gallium arsenide wafer entering the processing facility can be expected to cost 10 to 20 times as much as a silicon wafer, although that cost difference narrows somewhat after fabrication, packaging, and testing. Nevertheless, there is one major characteristic of III–V materials with which silicon cannot compete: a III–V compound can be tailored to generate or detect photons of a specific wavelength. For example, an indium gallium arsenide phosphide (InGaAsP) laser can generate radiation at 1.55 micrometres to carry digitally coded information streams. This means that a III–V component can fill both electronic and photonic functions in the same integrated circuit.

Photoresist Films

Patterning polished wafers with an integrated circuit requires the use of photoresist materials that form thin coatings on the wafer before each step of the photolithographic process. Modern photoresists are polymeric materials that are modified when exposed to radiation (either in the form of visible, ultraviolet, or X-ray photons or in the form of energetic electron beams). A photoresist typically contains a photoactive compound (PAC) and an alkaline-soluble resin. The PAC, mixed into the resin, renders it insoluble. This mixture is coated onto the semiconductor wafer and is then exposed to radiation through a "mask" that carries the desired pattern. Exposed PAC is converted into an acid that renders the resin soluble, so that the resist can be dissolved and the exposed substrate beneath it chemically etched or metallically coated to match the circuit design.

Besides practical properties such as shelf life, cost, and availability, the key properties of a photoresist include purity, etching resistance, resolution, contrast, and sensitivity. As the feature sizes of integrated circuits shrink in each successive generation of microchips, photoresist materials are challenged to handle shorter wavelengths of light. For example, the photolithography of current designs (with features that have shrunk to less than one micrometre) is based on ultraviolet radiation in the wavelength range of 365 to 436 nanometres, but, in order to define accurately the smaller features of future microchips (less than 0.25 micrometre), shorter wavelengths will be necessary. The problem here is that electromagnetic radiation in such frequency regions is weaker. One solution is to use the chemically amplified photoresist, or CAMP. The sensitivity of a photoresist is measured by its quantum efficiency, or the number of chemical events that occur when a photon is absorbed by the material. In CAMP material, the number of events is dramatically increased by subsequent chemical reactions (hence the amplification), which means that less light is needed to complete the process.

Electric Connections

The performance of today's electronic systems (and photonic systems as well) is limited significantly by interconnection technology, in which components and subsystems are linked by conductors and connectors. Currently, very fine gold or copper wiring, as thin as 30 micrometres, is used to carry electric current to and from the many pads along the sides or ends of a microchip to other components on a circuit board. The capacitance involved in such circuitry slows down the flow of electrons and, hence, of information. However, by integrating several chips into a single multichip module, in which the chips are connected on a shared substrate by various conducting materials (such as metalized film), the speed of information flow can be increased, thus improving the assembly's performance. Ideally, all the chips in a single module would be fabricated simultaneously on the same wafer, but in practice this is not feasible: Silicon crystal manufacture is still subject to an average of one flaw per wafer, meaning that at least one of the many chips cut from each wafer is scrapped. If the whole wafer area were dedicated to a single multifunction assembly, that one flaw would scrap the entire module. Multichip modules are therefore made up of as many as five microchips bonded to a silicon or ceramic substrate on which resistors and capacitors have been constructed with thin films. Typical materials used in a multichip module include the substrate; gold paste conductors applied in an additive process resembling silk screen printing; vitreous glazes to insulate the gold paste conductors from subsequent film layers; a series of thin films made with tantalum nitride, titanium, palladium, and plated gold; and a final package of silicone rubber.

Packaging Materials

Several major types of packaging material are used by the electronics industry, including ceramic, refractory glass, premolded plastic, and postmolded plastic. Ceramic and glass packages cost more than plasticpackages, so they make up less than 10 percent of the worldwide total. However, they provide the best protection for complex chips. Premolded plastic packages account for only a small but important fraction of the market, since they are required for packaging devices with many leads. Most plastic packages are postmolded, meaning that the package body is molded over the assembly after the microchip has been attached to the fan-out pattern.

Precursors

The starting materials for most semiconductor devices are volatile and ultrapure gaseous derivatives of various organic and inorganic precursors. Many of them are toxic, and many will ignite spontaneously in the atmosphere. These gases are transported in high-pressure cylinders from the plant where they were made to the site where they will be used. One possible method of replacing these precursors with materials that are environmentally safe is known as in situ synthesis. In this method, dangerous reagents would be generated on demand in only the desired quantities, instead of being shipped cross-country and stored until needed at the semiconductor processing plant.

Photonic Materials

Computers and communications systems have been dominated by electronic technology since their beginnings, but photonic technology is making serious inroads throughout the information movement and management systems with such devices as lasers, light-emitting diodes, photodetecting diodes, optical switches, optical amplifiers, optical modulators, and optical fibres. Indeed, for long-distance terrestrial and transoceanic transmission of information, photonics has almost completely displaced electronics.

Crystalline Materials

The light detectors and generators are actually optoelectronic, because they link photonic and electronic systems. They employ the III–V compound semiconductors, many of them characterized by their band gaps—*i.e.,* the energy minimum of the electron conduction band and the energy maximum of hole valence bands occur at the same location in the momentum space, allowing electrons and holes to recombine and radiate photons efficiently. (By contrast, the conduction band minimum and the valence band maximum in silicon have dissimilar momenta, and therefore the electrons and holes cannot recombine efficiently.) Among the important compounds are gallium arsenide, aluminum gallium arsenide, indium gallium arsenide phosphide, indium phosphide, and aluminum indium arsenide.

Fabricating a single crystal from these combinations of elements is far more difficult than creating a single crystal of electronic-grade silicon. Special furnaces are required, and the process can take several days. Notwithstanding the precision involved, the sausage-shaped boule is less than half the diameter of a silicon ingot and is subject to a much higher rate of defects. Researchers are continuously seeking ways to reduce the thermal stresses that are primarily responsible for dislocations in the III–V crystal lattice that cause these defects. The purity and structural perfection of the final single-crystal substrates affect the qualities of the crystalline layers that are grown on them and the regions that are diffused or implanted in them during the manufacture of photonic devices.

Epitaxial Layers

For the efficient emission or detection of photons, it is often necessary to constrain these processes to very thin semiconductor layers. These thin layers, grown atop bulk semiconductor wafers, are called epitaxial layers because their crystallinity matches that of the substrate even though the composition of the materials may differ—*e.g.*, gallium aluminum arsenide (GaAlAs) grown atop a gallium arsenide substrate. The resulting layers form what is called a heterostructure. Most continuously operating semiconductor lasers consist of heterostructures, a simple example consisting of 1000-angstrom thick gallium arsenide layers sandwiched between somewhat thicker (about 10000 angstroms) layers of gallium aluminum arsenide—all grown epitaxially on a gallium arsenide substrate. The sandwiching and repeating of very thin layers of a semiconductor between layers of a different composition allow one to modify the band gap of the sandwiched layer. This technique, called band-gap engineering, permits the creation of semiconductor materials with properties that cannot be found in nature. Band-gap engineering, used extensively with III–V compound semiconductors, can also be applied to elemental semiconductors such as silicon and germanium.

The most precise method of growing epitaxial layers on a semiconducting substrate is molecular-beam epitaxy (MBE). In this technique, a stream or beam of atoms or molecules is effused from a common source and travels across a vacuum to strike a heated crystal surface, forming a layer that has the same crystal structure as the substrate. Variations of MBE include elemental-source MBE, hydride-source MBE, gas-source MBE, and metal-organic MBE. Other approaches to epitaxial growth are liquid-phase epitaxy (LPE) or chemical vapour deposition (CVD). The latter method includes hydride CVD, trichloride CVD, and metal-organic CVD.

Normally, epitaxial layers are grown on flat surfaces, but scientists are searching for an economical and reliable method of growing epitaxial material on nonplanar structures—for example, around the "mesas" or "ridges" or in the "tubs" or "channels" that are etched into the surface of semiconducting devices. Nonplanar epitaxy is considered necessary for producing monolithic integrated optical devices or all-photonic switches and logic elements, but mastery of this method requires better understanding of the surface chemistry and surface dynamics of epitaxial growth.

Optical Switching

Research in this area is driven by the need to switch data streams of higher and higher speed efficiently as customers for computer and communications services demand transmission and switching rates far higher than can be provided by a purely electronic system. Thanks to developments in semiconductor lasers and detectors and in optical fibres, transmission at the desired high speeds has become possible. However, the switching of optical data streams still requires converting the data from the optical to the electronic domain, subjecting them to electronic switching and to manipulation inside the switching apparatus, and then reconverting the switched and reconfigured data into the optical domain for transmission over optical fibres. Electronic switching therefore is seen as the principal barrier to achieving higher switching speeds. One approach to solving this problem would be to introduce optics inside digital switching machines. Known as free-space photonics, this approach would involve such devices as semiconductor lasers or light-emitting diodes (LEDs), optical modulators, and photodetectors—all of which would be integratedinto systems combined with electronic components.

One commercially available device for photonic switching is the quantum-well self-electro-optic-effect device, or SEED. The key concept for this device is the use of quantum wells. These structures consist of many thin layers of two different semiconductor materials. Individual layers are typically 10 nanometres (about 40 atoms) thick, and 100 layers are used in a device about 1 micrometre thick. When a voltage is applied across the layers, the transmission of photons through the quantum wells changes significantly, in effect creating an optical modulator—an essential component of any photonic circuit. Variations on the SEED concept are the symmetric SEED (S-SEED) and the field-effect transistor SEED. Neighbouring S-SEEDs could be connected by pairs of back-to-back quantum-well photodiodes, and commercially sized interconnection networks could be built by using free-space photonic interconnections between two-dimensional arrays of switching nodes. However, even this type of free-space optical interconnection technology would only enhanceand extend electronic technology, not replace it.

The move of optoelectronic and photonic integrated circuits out of the research laboratory and into the marketplace has been made possible by the availability of high-quality epitaxial growth techniques for building up lattice-matched crystalline layers of indium gallium arsenide phosphide and indium phosphide (InGaAsP/InP). This III–V compound system is central to the light emitters and detectors used in the 1.3-micrometre and 1.5-micrometre wavelength ranges at which optical fibre has very low transmission loss.

Optical Transmission

As the rates of transmission are increased from millions of bits (megabits) per second to billions of bits (gigabits) per second, commercially available lasers encounter a physical limitation called "chirping," in which the optical frequency of the laser begins to waver during a pulse. Future systems, which may require from 2.4 to 30 gigabits per second, are probably going to be based on the use of a continuously operating distributed-feedback laser, whose output will be modulated in intensity by passing it through a modulator. This device consists of a crystal substrate of lithium niobate onto which a titanium channel is diffused to function as a light guide. The signal is encoded onto the light beam via a microwave radio-frequency feed through neighbouring channels in the coupler. Such a device is used only at the transmitter end of the optical path.

Both communications and computer systems rely on silica glass fibres to transmit light signals from lasers and LEDs. For long-distance transmission, optical-fibre cables are usually equipped with electro-optical repeater assemblies approximately every 100 kilometres. A new approach, called optical amplifiers, has been developed for deployment in transoceanicfibre-optic cables. Unlike traditional repeaters, optical amplifiers work by adding photons to a light signal without changing it to an electrical signal and without changing its bit-rate. Since they can be used at any desired transmission bit-rate, a transoceanic cable equipped with these devices can be upgraded to higher bit-rates simply by changing the lasers and photodiodes at each end. No retrofitting of higher bit-rate amplifiers is necessary.

The optical amplifier is a module containing a semiconductor pump laser and a short length of optical fibre whose core has been doped with less than 0.1 percent erbium, an optically active rare-earth element. The pump laser is powered by an electrical conductor that runs the length of the cable. The amplifier functions by converting the optical energy generated by the pump source into signal photon energy. When a signal-carrying stream of laser pulses passes through the optical amplifier, it is combined with the pump light through a wavelength division multiplexer located in the module. The combined signal is fed through the erbium-doped fibre length, where the excited erbium ions contribute photons coherently to the signal. The amplified signal is then fed to the next section of cable for transmission to the next optical amplifier, perhaps 200 to 300 kilometres away.

Materials for Medicine

The treatment of many human disease conditions requires surgical intervention in order to assist, augment, sustain, or replace a diseased organ, and such procedures involve the use of materials foreign to the body. These materials, known as biomaterials, include synthetic polymers and, to a lesser extent, biological polymers, metals, and ceramics. Specific applications of biomaterials range from high-volume products such as blood bags, syringes, and needles to more challenging implantable devices designed to augment or replace a diseased human organ. The latter devices are used in cardiovascular, orthopedic, and dental applications as well as in a wide range of invasive treatment and diagnostic systems. Many of these devices have made possible notable clinical successes. For example, in cardiovascular applications, thousands of lives have been saved by heart valves, heart pacemakers, and large-diameter vascular grafts, and orthopedic hip-joint replacements have shown great long-term success in the treatment of patients suffering from debilitating joint diseases. With such a tremendous increase in medical applications, demand for a wide range of biomaterials grows by 5 to 15 percent each year. In the United States the annual market for surgical implants exceeds $10 billion, approximately 10 percent of world demand.

Nevertheless, applications of biomaterials are limited by biocompatibility, the problem of adverse interactions arising at the junction between the biomaterial and the host tissue. Optimizing the interactions that occur at the surface of implanted biomaterials represents the most significant key to further advances, and an excellent basis for these advances can be found in the growing understanding of complex biological materials and in the development of novel biomaterials custom-designed at the molecular level for specific medical applications.

General Requirements of Biomaterials

Research on developing new biomaterials is an interdisciplinary effort, often involving collaboration among materials scientists and engineers, biomedical engineers, pathologists, and clinicians to solve clinical problems. The design or selection of a specific biomaterial depends on the relative importance of the various properties that are required for the intended medical application. Physical properties that are generally considered include hardness, tensile strength, modulus, and elongation; fatigue strength, which is determined by a material's response to cyclic loads or strains; impact properties; resistance to abrasion and wear; long-term dimensional stability, which is described by a material's viscoelastic properties; swelling in aqueous media; and permeability to gases, water, and small biomolecules. In addition, biomaterials are exposed to human tissues and fluids, so that predicting the results of possible interactions between host and material is an important and unique consideration in using synthetic materials in medicine. Two particularly important issues in biocompatibility are thrombosis, which involves blood coagulation and the adhesion of blood platelets to biomaterial surfaces, and the fibrous-tissue encapsulation of biomaterials that are implanted in soft tissues.

Poor selection of materials can lead to clinical problems. One example of this situation was the choice of silicone rubber as a poppet in an early heart valve design. The silicone absorbed lipid from plasma and swelled sufficiently to become trapped between the metal struts of the valve. Another unfortunate choice as a biomaterial was Teflon (trademark), which is noted for its low coefficient of friction and its chemical inertness but which has relatively poor abrasion resistance. Thus, as an occluder in a heart valve or as an acetabular cup in a hip-joint prosthesis, Teflon may eventually wear to such an extent that the device would fail. In addition, degradable polyesterurethane foam was abandoned as a fixation patch for breast prostheses, because it offered a distinct possibility for the release of carcinogenic by-products as it degraded.

Besides their constituent polymer molecules, synthetic biomaterials may contain several additives, such as unreacted monomers and catalysts, inorganic fillers or organic plasticizers, antioxidants and stabilizers, and processing lubricants or mold-release agents on the material's surface. In addition, several degradation products may result from the processing, sterilization, storage, and ultimately implantation of a device. Many additives are beneficial—for example, the silica filler that is indispensable in silicone rubber for good mechanical performance or the antioxidants and stabilizers that prevent premature oxidative degradation of polyetherurethanes. Other additives, such as pigments, can be eliminated from biomedical products. Indeed, a "medical-grade" biomaterial is one that has had nonessential additives and potential contaminants excluded or eliminated from the polymer. In order to achieve this grade, the polymer may need to be solvent-extracted before use, thereby eliminating low-molecular-weight materials. Generally, additives in polymers are regarded with extreme suspicion, because it is often the additives rather than the constituent polymer molecules that are the source of adverse biocompatibility.

Polymer Biomaterials

The majority of biomaterials used in humans are synthetic polymers such as the polyurethanes or dacron, rather than polymers of biological origin such as proteins or polysaccharides. The properties of common synthetic biomaterials vary widely, from the soft and delicate water-absorbing hydrogels made into contact lenses to the resilient elastomers found in short- and long-term cardiovascular devices or the high-strength acrylics used in orthopedics and dentistry. The properties of any material are governed by its chemical composition and by the intra- and intermolecular forces that dictate its molecular organization. Macromolecular structure in turn affects macroscopic properties and, ultimately, the interfacial behaviour of the material in contact with blood or host tissues.

Since the properties of each material are dependent on the chemical structure and macromolecular organization of its polymer chains, an understanding of some common structural features of various polymers provides considerable insight into their properties. Compared with complex biological molecules, synthetic polymers are relatively simple; often they comprise only one type of repeating subunit, analogous to a polypeptide consisting of just one repeating amino acid. On the basis of common structures and properties, synthetic polymers are classified into one of three categories: elastomers, which include natural and synthetic rubbers; thermoplastics; and thermosets. The properties that provide the basis for this classification include molecular weight, cross-link density, percent crystallinity, thermal transition temperature, and bulk mechanical properties.

Elastomers

Elastomers, which include rubber materials, have found wide use as biomaterials in cardiovascular and soft-tissue applications owing to their high elasticity, impact resistance, and gas permeability. Applications of elastomers include flexible tubing for pacemaker leads, vascular grafts, and catheters; biocompatible coatings and pumping diaphragms for artificial hearts and left-ventricular assist devices; grafts for reconstructive surgery and maxillofacial operations; wound dressings; breast prostheses; and membranes for implantable biosensors.

Elastomers are typically amorphous with low cross-link density (although linear polyurethane block copolymers are an important exception). This gives them low to moderate modulus and tensile properties as well as high elasticity. For example, elastomeric devices can be extended by 100 to 1,000 percent of their initial dimensions without causing any permanent deformation to the material. Silicone rubbers such as Silastic (trademark), produced by the American manufacturer Dow Corning, Inc., are cross-linked, so that they cannot be melted or dissolved—although swelling may occur in the presence of a good solvent. Such properties contrast with those of the linear polyurethane elastomers, which consist of soft polyether amorphous segments and hard urethane-containing glassy or crystalline segments. The two segments are incompatible at room temperature and undergo microphase separation, forming hard domains dispersed in an amorphous matrix. A key feature of this macromolecular organization is that the hard domains serve as physical cross-links and reinforcing filler. This results in elastomeric materials that possess relatively high modulus and extraordinary long-term stability under sustained cyclic loading. In addition, they can be processed by methods common to thermoplastics.

Thermoplastics

Many common thermoplastics, such as polyethylene and polyester, are used as biomaterials. Thermoplastics usually exhibit moderate to high tensile strength (5 to 1,000 megapascals) with moderate elongation (2 to 100 percent), and they undergo plastic deformation at high strains. Thermoplastics consist of linear or branched polymer chains; consequently, most can undergo reversible melt-solid transformation on heating, which allows for relatively easy processing or reprocessing. Depending on the structure and molecular organization of the polymer chains, thermoplastics may be amorphous (*e.g.*, polystyrene), semicrystalline (*e.g.*, low-density polyethylene), or highly crystalline (*e.g.*, high-density polyethylene), or they may be processed into highly crystalline textile fibres (*e.g.*, polyester Dacron).

Some thermoplastic biomaterials, such as polylactic acid and polyglycolic acid, are polymers based on a repeating amino acid subunit. These polypeptides are biodegradable, and, along with biodegradable polyesters and polyorthoesters, they have applications in absorbable sutures and drug-release systems. The rate of biodegradation in the body can be adjusted by using copolymers. These are polymers that link two different monomer subunits into a single polymer chain. The resultant biomaterial exhibits properties, including biodegradation, that are intermediate between the two homopolymers.

Thermosets

Thermosetting polymers find only limited application in medicine, but their characteristic properties, which combine high strength and chemical resistance, are useful for some orthopedic and dental devices. Thermosetting polymers such as epoxies and acrylics are chemically inert, and they also have high modulus and tensile properties with negligible elongation (1 to 2 percent). The polymer chains in these materials are highly cross-linked and therefore have severely restricted macromolecular mobility; this limits extension of the polymer chains under an applied load. As a result, thermosets are strong but brittle materials.

Cross-linking inhibits close packing of polymer chains, preventing formation of crystalline regions. Another consequence of extensive cross-linking is that thermosets do not undergo solid-melt transformation on heating, so that they cannot be melted or reprocessed.

Applications of Biomaterials

Cardiovascular Devices

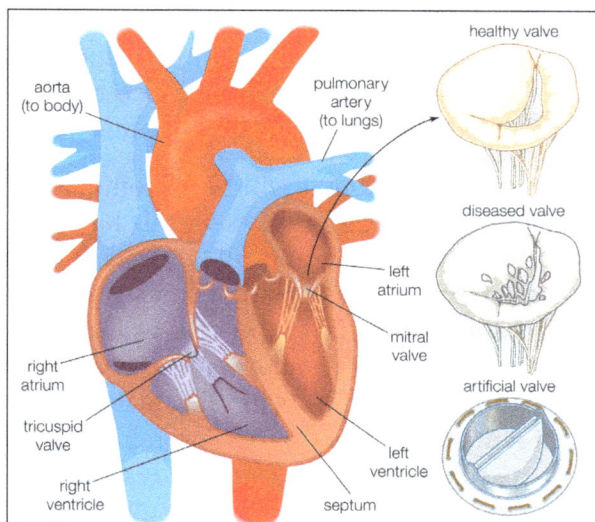

Diagram showing a normal heart valve compared with an artificial heart valve.

Biomaterials are used in many blood-contacting devices. These include artificial heart valves, synthetic vascular grafts, ventricular assist devices, drug-release systems, extracorporeal systems, and a wide range of invasive treatment and diagnostic systems. An important issue in the design and selection of materials is the hemodynamic conditions in the vicinity of the device. For example, mechanical heart valve implants are intended for long-term use. Consequently, the hinge points of each valve leaflet and the materials must have excellent wear and fatigue resistance in order to open and close 80 times per minute for many years after implantation. In addition, the open valve must minimize disturbances to blood flow as blood passes from the left ventricle of the heart, through the heart valve, and into the ascending aorta of the arterial vascular system. To this end, the bileaflet valve disks of one type of implant are coated with pyrolytic carbon, which provides a relatively smooth, chemically inert surface. This is an important property, because surface roughness will cause turbulence in the blood flow, which in turn may lead to hemolysis of red cells, provide sites for adventitious bacterial adhesion and subsequent colonization, and, in areas of blood stasis, promote thrombosis and blood coagulation.

The carbon-coated holding ring of this implant is covered with Dacron mesh fabric so that the surgeon can sew and fix the device to adjacent cardiac tissues. Furthermore, the porous structure of the Dacron mesh promotes tissue integration, which occurs over a period of weeks after implantation.

While the possibility of thrombosis can be minimized in blood-contacting biomaterials, it cannot be eliminated entirely. For this reason, patients who receive artificial heart valves or other blood-contacting devices also receive anticoagulation therapy. This is needed because all foreign surfaces initiate blood coagulation and platelet adhesion to some extent. Platelets are circulating cellular components of blood, two to four micrometres in size, that attach to foreign surfaces and actively participate in blood coagulation and thrombus formation. Research on new biomaterials for cardiovascular applications is largely devoted to understanding thrombus formation and to developing novel surfaces for biomaterials that will provide improved blood compatibility.

Synthetic vascular graft materials are used to patch injured or diseased areas of arteries, for replacement of whole segments of larger arteries such as the aorta, and for use as sewing cuffs. Such materials need to be flexible to allow for the difficulties of implantation and to avoid irritating adjacent tissues; also, the internal diameter of the graft should remain constant under a wide range of flexing and bending conditions, and the modulus or compliance of the vessel should be similar to that of the natural vessel. These aims are largely achieved by crimped woven Dacron and expanded polytetrafluoroethylene (ePTFE). Crimping of Dacron in processing results in a porous vascular graft that may be bent 180° or twisted without collapsing the internal diameter.

A biomaterial used for blood vessel replacement will be in contact not only with blood but also with adjacent soft tissues. Experience with different materials has shown that tissue growth into the interstices of the biomaterials aids healing and integration of the material with host tissue after implantation. In order for the tissue, which consists mostly of collagen, to grow in the graft, the vascular graft must have an open structure with pores at least 10 micrometres in diameter. These pores allow new blood capillaries that develop during healing to grow into the graft, and the blood then provides oxygen and other nutrients for fibroblasts and other cells to survive in the biomaterial matrix. Fibroblasts synthesize the structural protein tropocollagen, which is needed in the development of new fibrous tissue as part of the healing response to a surgical wound.

Occasionally, excessive tissue growth may be observed at the anastomosis, which is where the graft is sewn to the native artery. This is referred to as internal hyperplasia and is thought to result from differences in compliance between the graft and the host vessels. In addition, in order to optimize compatibility of the biomaterial with the blood, the synthetic graft eventually should be coated with a confluent layer of host endothelial cells, but this does not occur with current materials. Therefore, most proposed modifications to existing graft materials involve potential improvements in blood compatibility.

Artificial heart valves and vascular grafts, while not ideal, have been used successfully and have saved many thousands of lives. However, the risk of thrombosis has limited the success of existing cardiovascular devices and has restricted potential application of the biomaterials to other devices. For example, there is an urgent clinical need for blood-compatible, synthetic vascular grafts of small diameter in peripheral vascular surgery—e.g., in the legs—but this is currently impracticable with existing biomaterials because of the high risk of thrombotic occlusion. Similarly, progress with implantable miniature sensors, designed to measure a wide range of blood conditions continuously, has been impeded because of problems directly attributable to the failure of existing biomaterials. With such biocompatibility problems resolved, biomedical sensors would provide a very important contribution to medical diagnosis and monitoring. Considerable advances have been made in the ability to manipulate molecular architecture at the surfaces of materials by using chemisorbed or physisorbed monolayer films. Such progress in surface modification, combined with the development of nanoscale probes that permit examination at the molecular and submolecular level, provide a strong basis for optimism in the development of specialty biomaterials with improved blood compatibility.

Orthopedic Devices

Joint replacements, particularly at the hip, and bone fixation devices have become very successful applications of materials in medicine. The use of pins, plates, and screws for bone fixation to aid recovery of bone fractures has become routine, with the number of annual procedures approaching five million in the United States alone. In joint replacement, typical patients are age 55 or older and suffer from debilitating rheumatoid arthritis, osteoarthritis, or osteoporosis. Orthopedic surgeries for artificial joints exceed 1.5 million each year, with actual joint replacement accounting for about half of the procedures. A major focus of research is the development of new biomaterials for artificial joints intended for younger, more active patients.

Hip-joint replacements are principally used for structural support. Consequently, they are dominated by materials that possess high strength, such as metals, tough plastics, and reinforced polymer-matrix composites. In addition, biomaterials used for orthopedic applications must have high modulus, long-term dimensional stability, high fatigue resistance, long-term biostability, excellent abrasion resistance, and biocompatibility (*i.e.,*there should be no adverse tissue response to the implanted device). Early developments in this field used readily available materials such as stainless steels, but evidence of corrosion after implantation led to their replacement by more stable materials, particularly titanium alloys, cobalt-chromium-molybdenum alloys, and carbon fibre-reinforced polymer composites. A typical modern artificial hip consists of a nitrided and highly polished cobalt-chromium ball connected to a titanium alloy stem that is inserted into the femur and cemented into place by in situ polymerization of polymethylmethacrylate. The articulating component of the joint consists of an acetabular cup made of tough, creep-resistant, ultrahigh-molecular-weight polyethylene. Abrasion at the ball-and-cup interface can lead to the production of wear particles, which in turn can lead to significant inflammatory reaction by the host. Consequently, much research on the development of hip-joint materials has been devoted to optimizing the properties of the articulating components in order to eliminate surface wear. Other modifications include porous coatings made by sintering the metal surface or coatings of wire mesh or hydroxyapatite; these promote bone growth and integration between the implant and the host, eliminating the need for an acrylic bone cement.

While the strength of the biomaterials is important, another goal is to match the mechanical properties of the implant materials with those of the bone in order to provide a uniform distribution of stresses (load sharing). If a bone is loaded insufficiently, the stress distribution will be made asymmetric, and this will lead to adaptive remodeling with cortical thinning and increased porosity of the bone. Structure hierarchy and in the structure-property relationships of materials have been obtained from studies on biologic composite materials, and they are being translated into new classes of synthetic biomaterials. One development is carbon fibre-reinforced polymer-matrix composites. Typical matrix polymers include polysulfone and polyetheretherketones. The strength of these composites is lower than that of metals, but it more closely approximates that of bone.

2

Sub-disciplines of Materials Science

Materials science can be divided into various sub-disciplines. Some of them are ceramic engineering, crystallography, materials informatics, polymer engineering, polymer science and metallurgy. The chapter closely examines these sub-disciplines of materials science to provide an extensive understanding of the subject.

CERAMIC ENGINEERING

Ceramic engineering is a branch of engineering which deals with the science and technology of creating object from inorganic and non-metallic materials.

Ceramic engineering combines principles of chemistry, physics and engineering. Fiber-optic devices, microprocessors and solar panels are some examples of ceramic sciences applied to everyday life. The use of ceramic material is on the rise as it a low-cost and efficient material. Ceramic engineering is needed right from the production of ceramic teeth, bones, and other fibre optic cables used for surgery to ceramic superconductors, lasers, etc. As the applications of Ceramic material is expanded tremendously due to the recent advances in the field of medicine which include bio-ceramics and other, ceramic engineering is a booming field.

Modern Industry

Now a multibillion-dollar a year industry, ceramic engineering and research has established itself as an important field of science. Applications continue to expand as researchers develop new kinds of ceramics to serve different purposes:

- Zirconium dioxide ceramics are used in the manufacture of knives. The blade of the ceramic knife will stay sharp for much longer than that of a steel knife, although it is more brittle and can be snapped by dropping it on a hard surface.

- Ceramics such as alumina, boron carbide and silicon carbide have been used in bulletproof vests to repel small arms rifle fire. Such plates are known commonly as trauma plates. Similar material is used to protect cockpits of some military aircraft, because of the low weight of the material.

- Silicon nitride parts are used in ceramic ball bearings. Their higher hardness means that they are much less susceptible to wear and can offer more than triple lifetimes. They also deform less under load meaning they have less contact with the bearing retainer walls and can roll faster. In very high speed applications, heat from friction during rolling can cause problems for metal bearings; problems which are reduced by the use of ceramics. Ceramics are also more chemically resistant and can be used in wet environments where steel bearings would rust. The major drawback to using ceramics is a significantly higher cost. In many cases their electrically insulating properties may also be valuable in bearings.

- In the early 1980s, Toyota researched production of an adiabatic ceramic engine which can run at a temperature of over 6000 °F (3300 °C). Ceramic engines do not require a cooling system and hence allow a major weight reduction and therefore greater fuel efficiency. Fuel efficiency of the engine is also higher at high temperature, as shown by Carnot's theorem. In a conventional metallic engine, much of the energy released from the fuel must be dissipated as waste heat in order to prevent a meltdown of the metallic parts. Despite all of these desirable properties, such engines are not in production because the manufacturing of ceramic parts in the requisite precision and durability is difficult. Imperfection in the ceramic leads to cracks, which can lead to potentially dangerous equipment failure. Such engines are possible in laboratory settings, but mass-production is not feasible with current technology.

- Work is being done in developing ceramic parts for gas turbine engines. Currently, even blades made of advanced metal alloys used in the engines' hot section require cooling and careful limiting of operating temperatures. Turbine engines made with ceramics could operate more efficiently, giving aircraft greater range and payload for a set amount of fuel.

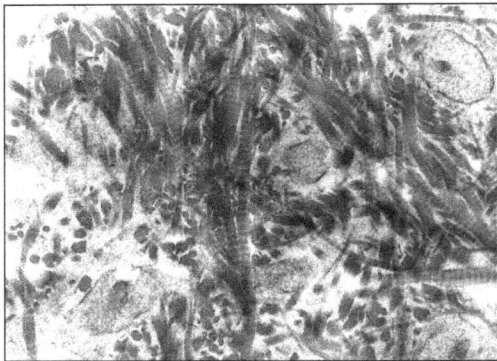

Collagen fibers of woven bone.

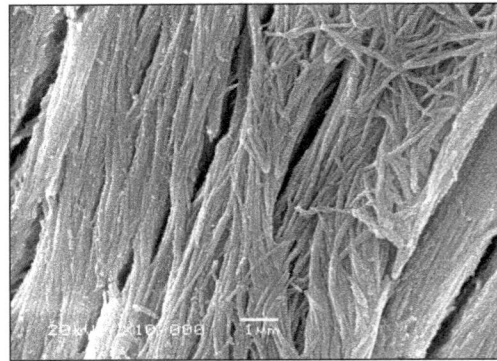

Scanning electron microscopy image of bone.

- Recently, there have been advances in ceramics which include bio-ceramics, such as dental implants and synthetic bones. Hydroxyapatite, the natural mineral component of bone, has been made synthetically from a number of biological and chemical sources and can be formed into ceramic materials. Orthopaedic implants made from these materials bond readily to bone and other tissues in the body without rejection or inflammatory reactions. Because of this, they are of great interest for gene delivery and tissue engineering scaffolds. Most hydroxyapatite ceramics are very porous and lack mechanical strength and are used to coat metal orthopaedic devices to aid in forming a bond to bone or as bone fillers. They are also used as fillers for orthopaedic plastic screws to aid in reducing the inflammation and increase absorption of these plastic materials. Work is being done to make strong, fully dense nano crystalline hydroxyapatite ceramic materials for orthopaedic weight bearing devices, replacing foreign metal and plastic orthopaedic materials with a synthetic, but naturally occurring, bone mineral. Ultimately these ceramic materials may be used as bone replacements or with the incorporation of protein collagens, synthetic bones.

- Durable actinide-containing ceramic materials have many applications such as in nuclear fuels for burning excess Pu and in chemically-inert sources of alpha irradiation for power supply of unmanned space vehicles or to produce electricity for microelectronic devices. Both use and disposal of radioactive actinides require their immobilisation in a durable host material. Nuclear waste long-lived radionuclides such as actinides are immobilised using chemically-durable crystalline materials based on polycrystalline ceramics and large single crystals.

Glass-ceramics

A high strength glass-ceramic cook-top with
negligible thermal expansion.

Glass-ceramic materials share many properties with both glasses and ceramics. Glass-ceramics have an amorphous phase and one or more crystalline phases and are produced by a so-called "controlled crystallization", which is typically avoided in glass manufacturing. Glass-ceramics often contain a crystalline phase which constitutes anywhere from 30% (m/m) to 90% (m/m) of its composition by volume, yielding an array of materials with interesting thermomechanical properties.

In the processing of glass-ceramics, molten glass is cooled down gradually before reheating and annealing. In this heat treatment the glass partly crystallizes. In many cases, so-called 'nucleation agents' are added in order to regulate and control the crystallization process. Because there is usually no pressing and sintering, glass-ceramics do not contain the volume fraction of porosity typically present in sintered ceramics.

The term mainly refers to a mix of lithium and aluminosilicates which yields an array of materials with interesting thermomechanical properties. The most commercially important of these have the distinction of being impervious to thermal shock. Thus, glass-ceramics have become extremely useful for countertop cooking. The negative thermal expansion coefficient (TEC) of the crystalline ceramic phase can be balanced with the positive TEC of the glassy phase. At a certain point (~70% crystalline) the glass-ceramic has a net TEC near zero. This type of glass-ceramic exhibits excellent mechanical properties and can sustain repeated and quick temperature changes up to 1000 °C.

Processing Steps

The traditional ceramic process generally follows this sequence: Milling → Batching → Mixing → Forming → Drying → Firing → Assembly.

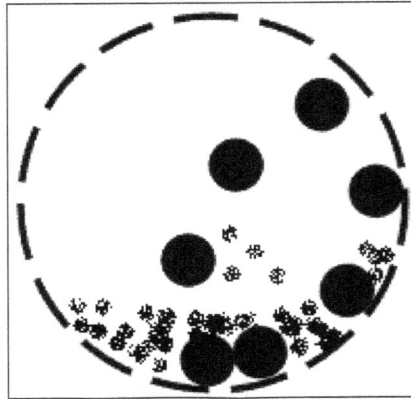

Ball mill.

- Milling is the process by which materials are reduced from a large size to a smaller size. Milling may involve breaking up cemented material (in which case individual particles retain their shape) or pulverization (which involves grinding the particles themselves to a smaller size). Milling is generally done by mechanical means, including *attrition* (which is particle-to-particle collision that results in agglomerate break up or particle shearing), *compression* (which applies a forces that results in fracturing), and *impact* (which employs a milling medium or the particles themselves to cause fracturing). Attrition milling equipment includes the wet scrubber (also called the planetary mill or wet attrition mill), which has paddles in water creating vortexes in which the material collides and break up. Compression mills include the jaw crusher, roller crusher and cone crusher. Impact mills include the ball mill, which has media that tumble and fracture the material. Shaft impactors cause particle-to particle attrition and compression.

- Batching is the process of weighing the oxides according to recipes, and preparing them for mixing and drying.

- Mixing occurs after batching and is performed with various machines, such as dry mixing ribbon mixers (a type of cement mixer), Mueller mixers, and pug mills. Wet mixing generally involves the same equipment.

- Forming is making the mixed material into shapes, ranging from toilet bowls to spark plug insulators. Forming can involve: (1) Extrusion, such as extruding "slugs" to make bricks, (2) Pressing to make shaped parts, (3) Slip casting, as in making toilet bowls, wash basins and ornamentals like ceramic statues. Forming produces a "green" part, ready for drying. Green parts are soft, pliable, and over time will lose shape. Handling the green product will change its shape. For example, a green brick can be "squeezed", and after squeezing it will stay that way.

- Drying is removing the water or binder from the formed material. Spray drying is widely used to prepare powder for pressing operations. Other dryers are tunnel dryers and periodic dryers. Controlled heat is applied in this two-stage process. First, heat removes water. This step needs careful control, as rapid heating causes cracks and surface defects. The dried part is smaller than the green part, and is brittle, necessitating careful handling, since a small impact will cause crumbling and breaking.

- Sintering is where the dried parts pass through a controlled heating process, and the oxides are chemically changed to cause bonding and densification. The fired part will be smaller than the dried part.

Forming Methods

Ceramic forming techniques include throwing, slipcasting, tape casting, freeze-casting, injection moulding, dry pressing, isostatic pressing, hot isostatic pressing (HIP), 3D printing and others. Methods for forming ceramic powders into complex shapes are desirable in many areas of technology. Such methods are required for producing advanced, high-temperature structural parts such as heat engine components and turbines. Materials other than ceramics which are used in these processes may include: wood, metal, water, plaster and epoxy—most of which will be eliminated upon firing.

These forming techniques are well known for providing tools and other components with dimensional stability, surface quality, high (near theoretical) density and microstructural uniformity. The increasing use and diversity of speciality forms of ceramics adds to the diversity of process technologies to be used.

Thus, reinforcing fibres and filaments are mainly made by polymer, sol-gel, or CVD processes, but melt processing also has applicability. The most widely used speciality form is layered structures, with tape casting for electronic substrates and packages being pre-eminent. Photo-lithography is of increasing interest for precise patterning of conductors and other components for such packaging. Tape casting or forming processes are also of increasing interest for other applications, ranging from open structures such as fuel cells to ceramic composites.

The other major layer structure is coating, where melt spraying is very important, but chemical and physical vapour deposition and chemical (e.g., sol-gel and polymer pyrolysis) methods are all seeing increased use. Besides open structures from formed tape, extruded structures, such as honeycomb catalyst supports, and highly porous structures, including various foams, for example, reticulated foam, are of increasing use.

Densification of consolidated powder bodies continues to be achieved predominantly by (pressureless) sintering. However, the use of pressure sintering by hot pressing is increasing, especially for non-oxides and parts of simple shapes where higher quality (mainly microstructural homogeneity) is needed, and larger size or multiple parts per pressing can be an advantage.

The Sintering Process

The principles of sintering-based methods are simple ("sinter" has roots in the English "cinder"). The firing is done at a temperature below the melting point of the ceramic. Once a roughly-held-to-gether object called a "green body" is made, it is baked in a kiln, where atomic and molecular diffusion processes give rise to significant changes in the primary microstructural features. This includes the gradual elimination of porosity, which is typically accompanied by a net shrinkage and overall densification of the component. Thus, the pores in the object may close up, resulting in a denser product of significantly greater strength and fracture toughness.

Another major change in the body during the firing or sintering process will be the establishment of the polycrystalline nature of the solid. This change will introduce some form of grain size distribution, which will have a significant impact on the ultimate physical properties of the material. The grain sizes will either be associated with the initial particle size, or possibly the sizes of aggregates or particle clusters which arise during the initial stages of processing.

The ultimate microstructure (and thus the physical properties) of the final product will be limited by and subject to the form of the structural template or precursor which is created in the initial stages of chemical synthesis and physical forming. Hence the importance of chemical powder and polymer processing as it pertains to the synthesis of industrial ceramics, glasses and glass-ceramics.

There are numerous possible refinements of the sintering process. Some of the most common involve pressing the green body to give the densification a head start and reduce the sintering time needed. Sometimes organic binders such as polyvinyl alcohol are added to hold the green body together; these burn out during the firing (at 200–350 °C). Sometimes organic lubricants are added during pressing to increase densification. It is common to combine these, and add binders and lubricants to a powder, then press. (The formulation of these organic chemical additives is an art in itself. This is particularly important in the manufacture of high performance ceramics such as those used by the billions for electronics, in capacitors, inductors, sensors, etc).

A slurry can be used in place of a powder, and then cast into a desired shape, dried and then sintered. Indeed, traditional pottery is done with this type of method, using a plastic mixture worked with the hands. If a mixture of different materials is used together in a ceramic, the sintering temperature is sometimes above the melting point of one minor component – a liquid phase sintering. This results in shorter sintering times compared to solid state sintering.

Strength of Ceramics

A material's strength is dependent on its microstructure. The engineering processes to which a material is subjected can alter its microstructure. The variety of strengthening mechanisms that alter the strength of a material include the mechanism of grain boundary strengthening. Thus, although yield strength is maximized with decreasing grain size, ultimately, very small grain sizes make the material brittle. Considered in tandem with the fact that the yield strength is the parameter that predicts plastic deformation in the material, one can make informed decisions on how to increase the strength of a material depending on its microstructural properties and the desired end effect.

The relation between yield stress and grain size is described mathematically by the Hall-Petch equation which is:

$$\sigma_y = \sigma_0 + \frac{k_y}{\sqrt{d}}$$

where k_y is the strengthening coefficient (a constant unique to each material), σ_o is a materials constant for the starting stress for dislocation movement (or the resistance of the lattice to dislocation motion), d is the grain diameter, and σ_y is the yield stress.

Theoretically, a material could be made infinitely strong if the grains are made infinitely small. This is, unfortunately, impossible because the lower limit of grain size is a single unit cell of the material. Even then, if the grains of a material are the size of a single unit cell, then the material is in fact amorphous, not crystalline, since there is no long range order, and dislocations can not be defined in an amorphous material. It has been observed experimentally that the microstructure with the highest yield strength is a grain size of about 10 nanometres, because grains smaller than this undergo another yielding mechanism, grain boundary sliding. Producing engineering materials with this ideal grain size is difficult because of the limitations of initial particle sizes inherent to nanomaterials and nanotechnology.

Theory of Chemical Processing

Microstructural Uniformity

In the processing of fine ceramics, the irregular particle sizes and shapes in a typical powder often lead to non-uniform packing morphologies that result in packing density variations in the powder compact. Uncontrolled agglomeration of powders due to attractive van der Waals forces can also give rise to in microstructural inhomogeneities.

Differential stresses that develop as a result of non-uniform drying shrinkage are directly related to the rate at which the solvent can be removed, and thus highly dependent upon the distribution of porosity. Such stresses have been associated with a plastic-to-brittle transition in consolidated bodies, and can yield to crack propagation in the unfired body if not relieved.

In addition, any fluctuations in packing density in the compact as it is prepared for the kiln are often amplified during the sintering process, yielding inhomogeneous densification. Some pores and other structural defects associated with density variations have been shown to play a detrimental role in the sintering process by growing and thus limiting end-point densities. Differential stresses arising from inhomogeneous densification have also been shown to result in the propagation of internal cracks, thus becoming the strength-controlling flaws.

It would therefore appear desirable to process a material in such a way that it is physically uniform with regard to the distribution of components and porosity, rather than using particle size distributions which will maximize the green density. The containment of a uniformly dispersed assembly of strongly interacting particles in suspension requires total control over particle-particle interactions. Monodisperse colloids provide this potential.

Monodisperse powders of colloidal silica, for example, may therefore be stabilized sufficiently to ensure a high degree of order in the colloidal crystal or polycrystalline colloidal solid which results from aggregation. The degree of order appears to be limited by the time and space allowed for longer-range correlations to be established.

Such defective polycrystalline colloidal structures would appear to be the basic elements of sub-micrometer colloidal materials science, and, therefore, provide the first step in developing a more rigorous understanding of the mechanisms involved in microstructural evolution in inorganic systems such as polycrystalline ceramics.

Self-assembly

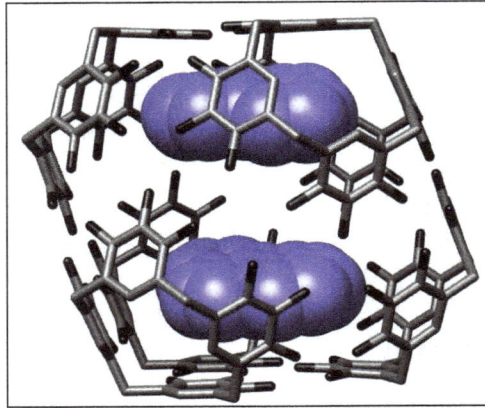

An example of a supramolecular assembly.

Self-assembly is the most common term in use in the modern scientific community to describe the spontaneous aggregation of particles (atoms, molecules, colloids, micelles, etc.) without the influence of any external forces. Large groups of such particles are known to assemble themselves into thermodynamically stable, structurally well-defined arrays, quite reminiscent of one of the 7 crystal systems found in metallurgy and mineralogy (e.g. face-centred cubic, body-centred cubic, etc.). The fundamental difference in equilibrium structure is in the spatial scale of the unit cell (or lattice parameter) in each particular case.

Thus, self-assembly is emerging as a new strategy in chemical synthesis and nanotechnology. Molecular self-assembly has been observed in various biological systems and underlies the formation of a wide variety of complex biological structures. Molecular crystals, liquid crystals, colloids, micelles, emulsions, phase-separated polymers, thin films and self-assembled monolayers all represent examples of the types of highly ordered structures which are obtained using these techniques. The distinguishing feature of these methods is self-organization in the absence of any external forces.

In addition, the principal mechanical characteristics and structures of biological ceramics, polymer composites, elastomers, and cellular materials are being re-evaluated, with an emphasis on bioinspired materials and structures. Traditional approaches focus on design methods of biological materials using conventional synthetic materials. This includes an emerging class of mechanically superior biomaterials based on microstructural features and designs found in nature. The new horizons have been identified in the synthesis of bioinspired materials through processes that are characteristic of biological systems in nature. This includes the nanoscale self-assembly of the components and the development of hierarchical structures.

Ceramic Composites

Substantial interest has arisen in recent years in fabricating ceramic composites. While there is considerable interest in composites with one or more non-ceramic constituents, the greatest attention is on composites in which all constituents are ceramic. These typically comprise two ceramic constituents: a continuous matrix, and a dispersed phase of ceramic particles, whiskers, or short (chopped) or continuous ceramic fibres. The challenge, as in wet chemical processing, is to obtain a uniform or homogeneous distribution of the dispersed particle or fibre phase.

The Porsche Carrera GT's carbon-ceramic
(silicon carbide) composite disc brake.

Consider first the processing of particulate composites. The particulate phase of greatest interest is tetragonal zirconia because of the toughening that can be achieved from the phase transformation from the metastable tetragonal to the monoclinic crystalline phase, aka transformation toughening. There is also substantial interest in dispersion of hard, non-oxide phases such as SiC, TiB, TiC, boron, carbon and especially oxide matrices like alumina and mullite. There is also interest too incorporating other ceramic particulates, especially those of highly anisotropic thermal expansion. Examples include Al_2O_3, TiO_2, graphite, and boron nitride.

Silicon carbide single crystal.

In processing particulate composites, the issue is not only homogeneity of the size and spatial distribution of the dispersed and matrix phases, but also control of the matrix grain size. However, there is some built-in self-control due to inhibition of matrix grain growth by the dispersed phase. Particulate composites, though generally offer increased resistance to damage, failure, or both, are still quite sensitive to inhomogeneities of composition as well as other processing defects such as pores. Thus they need good processing to be effective.

Particulate composites have been made on a commercial basis by simply mixing powders of the two constituents. Although this approach is inherently limited in the homogeneity that can be achieved, it is the most readily adaptable for existing ceramic production technology. However, other approaches are of interest.

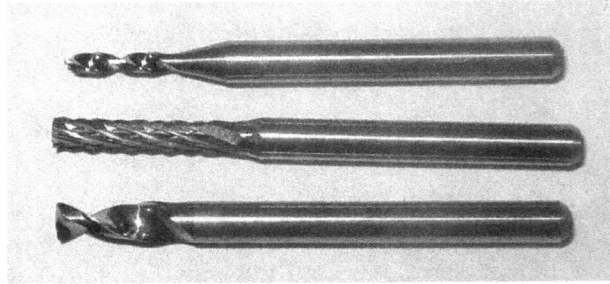

Tungsten carbide milling bits.

From the technological standpoint, a particularly desirable approach to fabricating particulate composites is to coat the matrix or its precursor onto fine particles of the dispersed phase with good control of the starting dispersed particle size and the resultant matrix coating thickness. One should in principle be able to achieve the ultimate in homogeneity of distribution and thereby optimize composite performance. This can also have other ramifications, such as allowing more useful composite performance to be achieved in a body having porosity, which might be desired for other factors, such as limiting thermal conductivity.

There are also some opportunities to utilize melt processing for fabrication of ceramic, particulate, whisker and short-fibre, and continuous-fibre composites. Clearly, both particulate and whisker composites are conceivable by solid-state precipitation after solidification of the melt. This can also be obtained in some cases by sintering, as for precipitation-toughened, partially stabilized zirconia. Similarly, it is known that one can directionally solidify ceramic eutectic mixtures and hence obtain uniaxially aligned fibre composites. Such composite processing has typically been limited to very simple shapes and thus suffers from serious economic problems due to high machining costs.

Clearly, there are possibilities of using melt casting for many of these approaches. Potentially even more desirable is using melt-derived particles. In this method, quenching is done in a solid solution or in a fine eutectic structure, in which the particles are then processed by more typical ceramic powder processing methods into a useful body. There have also been preliminary attempts to use melt spraying as a means of forming composites by introducing the dispersed particulate, whisker, or fibre phase in conjunction with the melt spraying process.

Other methods besides melt infiltration to manufacture ceramic composites with long fibre reinforcement are chemical vapour infiltration and the infiltration of fibre preforms with organic precursor, which after pyrolysis yield an amorphous ceramic matrix, initially with a low density. With repeated cycles of infiltration and pyrolysis one of those types of ceramic matrix composites is produced. Chemical vapour infiltration is used to manufacture carbon/carbon and silicon carbide reinforced with carbon or silicon carbide fibres.

Besides many process improvements, the first of two major needs for fibre composites is lower fibre costs. The second major need is fibre compositions or coatings, or composite processing, to reduce degradation that results from high-temperature composite exposure under oxidizing conditions.

Applications

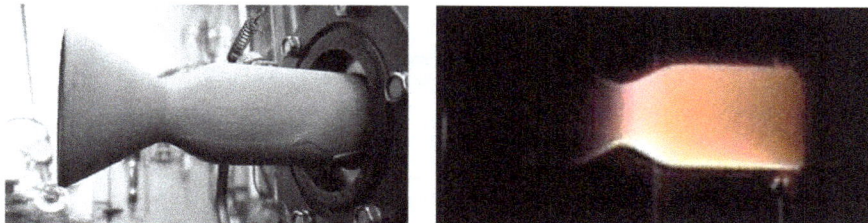

Silicon nitride thruster. Left: Mounted in test stand.
Right: Being tested with H_2/O_2 propellants.

The products of technical ceramics include tiles used in the Space Shuttle program, gas burner nozzles, ballistic protection, nuclear fuel uranium oxide pellets, bio-medical implants, jet engine turbine blades, and missile nose cones.

Its products are often made from materials other than clay, chosen for their particular physical properties. These may be classified as follows:

- Oxides: Silica, alumina, zirconia.

- Non-oxides: Carbides, borides, nitrides, silicides.

- Composites: Particulate or whisker reinforced matrices, combinations of oxides and non-oxides (e.g. polymers).

Ceramics can be used in many technological industries. One application is the ceramic tiles on NASA's Space Shuttle, used to protect it and the future supersonic space planes from the searing heat of re-entry into the Earth's atmosphere. They are also used widely in electronics and optics. In addition to the applications listed here, ceramics are also used as a coating in various engineering cases. An example would be a ceramic bearing coating over a titanium frame used for an aircraft. Recently the field has come to include the studies of single crystals or glass fibres, in addition to traditional polycrystalline materials, and the applications of these have been overlapping and changing rapidly.

Aerospace

- Engines: Shielding a hot running aircraft engine from damaging other components.

- Airframes: Used as a high-stress, high-temp and lightweight bearing and structural component.

- Missile nose-cones: Shielding the missile internals from heat.

- Space Shuttle tiles.

- Space-debris ballistic shields: ceramic fiber woven shields offer better protection to hyper-velocity (~7 km/s) particles than aluminium shields of equal weight.

- Rocket nozzles: Focusing high-temperature exhaust gases from the rocket booster.

- Unmanned Air Vehicles: Ceramic engine utilization in aeronautical applications (such as Unmanned Air Vehicles) may result in enhanced performance characteristics and less operational costs.

Biomedical

A titanium hip prosthesis, with a ceramic head
and polyethylene acetabular cup.

- Artificial bone; Dentistry applications, teeth

- Biodegradable splints; Reinforcing bones recovering from osteoporosis

- Implant material

Electronics

- Capacitors

- Integrated circuit packages

- Transducers

- Insulators

Optical

- Optical fibres, guided lightwave transmission

- Switches

- Laser amplifiers

- Lenses

- Infrared heat-seeking devices

Automotive

- Heat shield

- Exhaust heat management

Biomaterials

Silicification is quite common in the biological world and occurs in bacteria, single-celled organisms, plants, and animals (invertebrates and vertebrates). Crystalline minerals formed in such environment often show exceptional physical properties (e.g. strength, hardness, fracture toughness) and tend to form hierarchical structures that exhibit microstructural order over a range of length or spatial scales. The minerals are crystallized from an environment that is undersaturated with respect to silicon, and under conditions of neutral pH and low temperature (0–40 °C). Formation of the mineral may occur either within or outside of the cell wall of an organism, and specific biochemical reactions for mineral deposition exist that include lipids, proteins and carbohydrates.

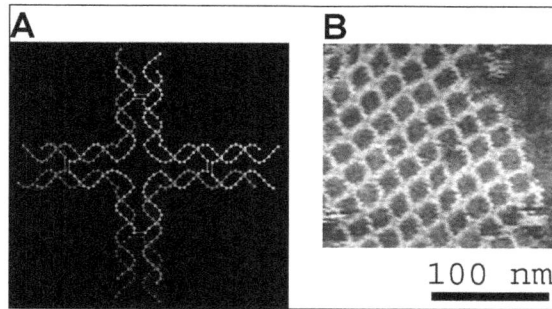

The DNA structure at left (schematic shown) will self-assemble into
the structure visualized by atomic force microscopy at right.

Most natural (or biological) materials are complex composites whose mechanical properties are often outstanding, considering the weak constituents from which they are assembled. These complex structures, which have risen from hundreds of million years of evolution, are inspiring the design of novel materials with exceptional physical properties for high performance in adverse conditions. Their defining characteristics such as hierarchy, multifunctionality, and the capacity for self-healing, are currently being investigated.

The basic building blocks begin with the 20 amino acids and proceed to polypeptides, polysaccharides, and polypeptides–saccharides. These, in turn, compose the basic proteins, which are the primary constituents of the 'soft tissues' common to most biominerals. With well over 1000 proteins possible, current research emphasizes the use of collagen, chitin, keratin, and elastin. The 'hard' phases are often strengthened by crystalline minerals, which nucleate and grow in a biomediated environment that determines the size, shape and distribution of individual crystals. The most important mineral phases have been identified as hydroxyapatite, silica, and aragonite. Using the classification of Wegst and Ashby, the principal mechanical characteristics and structures of biological ceramics, polymer composites, elastomers, and cellular materials have been presented. Selected systems in each class are being investigated with emphasis on the relationship between their microstructure over a range of length scales and their mechanical response.

Thus, the crystallization of inorganic materials in nature generally occurs at ambient temperature and pressure. Yet the vital organisms through which these minerals form are capable of consistently producing extremely precise and complex structures. Understanding the processes in which living organisms control the growth of crystalline minerals such as silica could lead to significant advances in the field of materials science, and open the door to novel synthesis techniques for nanoscale composite materials, or nanocomposites.

High-resolution SEM observations were performed of the microstructure of the mother-of-pearl (or nacre) portion of the abalone shell. Those shells exhibit the highest mechanical strength and fracture toughness of any non-metallic substance known. The nacre from the shell of the abalone has become one of the more intensively studied biological structures in materials science. Clearly visible in these images are the neatly stacked (or ordered) mineral tiles separated by thin organic sheets along with a macrostructure of larger periodic growth bands which collectively form what scientists are currently referring to as a hierarchical composite structure. (The term hierarchy simply implies that there are a range of structural features which exist over a wide range of length scales).

The iridescent nacre inside a Nautilus shell.

Future developments reside in the synthesis of bio-inspired materials through processing methods and strategies that are characteristic of biological systems. These involve nanoscale self-assembly of the components and the development of hierarchical structures.

CRYSTALLOGRAPHY

Crystallography is the experimental science of determining the arrangement of atoms in crystalline solids. X-ray crystallography is used to determine the structure of large biomolecules such as proteins. Before the development of X-ray diffraction crystallography, the study of crystals was based on physical measurements of their geometry. This involved measuring the angles of crystal faces relative to each other and to theoretical reference axes (crystallographic axes), and establishing the symmetry of the crystal in question. This physical measurement is carried out using a goniometer. The position in 3D space of each crystal face is plotted on a stereographic net such as a Wulff net or Lambert net. The pole to each face is plotted on the net. Each point is labelled with its Miller index. The final plot allows the symmetry of the crystal to be established.

Crystallographic methods now depend on analysis of the diffraction patterns of a sample targeted by a beam of some type. X-rays are most commonly used; other beams used include electrons or neutrons. This is facilitated by the wave properties of the particles. Crystallographers often explicitly state the type of beam used, as in the terms X-ray crystallography, neutron diffraction and electron diffraction. These three types of radiation interact with the specimen in different ways.

- X-rays interact with the spatial distribution of electrons in the sample.

- Electrons are charged particles and therefore interact with the total charge distribution of both the atomic nuclei and the electrons of the sample.

- Neutrons are scattered by the atomic nuclei through the strong nuclear forces, but in addition, the magnetic moment of neutrons is non-zero. They are therefore also scattered by magnetic fields. When neutrons are scattered from hydrogen-containing materials, they produce diffraction patterns with high noise levels. However, the material can sometimes be treated to substitute deuterium for hydrogen.

Because of these different forms of interaction, the three types of radiation are suitable for different crystallographic studies.

Theory

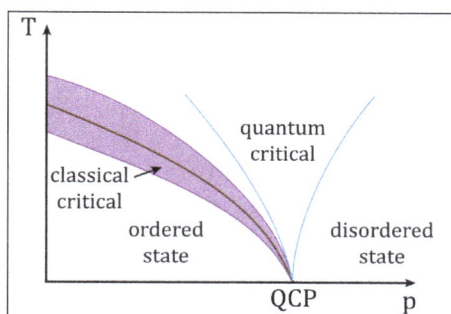

Condensed matter physics.

An image of a small object is made using a lens to focus the beam, similar to a lens in a microscope. However, the wavelength of visible light (about 4000 to 7000 ångström) is three orders of magnitude longer than the length of typical atomic bonds and atoms themselves (about 1 to 2 Å). Therefore, obtaining information about the spatial arrangement of atoms requires the use of radiation with shorter wavelengths, such as X-ray or neutron beams. Employing shorter wavelengths implied abandoning microscopy and true imaging, however, because there exists no material from which a lens capable of focusing this type of radiation can be created. Scientists have had some success focusing X-rays with microscopic Fresnel zone plates made from gold, and by critical-angle reflection inside long tapered capillaries. Diffracted X-ray or neutron beams cannot be focused to produce images, so the sample structure must be reconstructed from the diffraction pattern. Sharp features in the diffraction pattern arise from periodic, repeating structure in the sample, which are often very strong due to coherent reflection of many photons from many regularly spaced instances of similar structure, while non-periodic components of the structure result in diffuse (and usually weak) diffraction features - areas with a higher density and repetition of atom order tend to reflect more light toward one point in space when compared to those areas with fewer atoms and less repetition.

Because of their highly ordered and repetitive structure, crystals give diffraction patterns of sharp Bragg reflection spots, and are ideal for analyzing the structure of solids.

Notation

- Coordinates in square brackets such as denote a direction vector (in real space).

- Coordinates in angle brackets or chevrons such as <100> denote a family of directions which are related by symmetry operations. In the cubic crystal system for example, <100> would mean,, or the negative of any of those directions.

- Miller indices in parentheses such as (100) denote a plane of the crystal structure, and regular repetitions of that plane with a particular spacing. In the cubic system, the normal to the (hkl) plane is the direction [hkl], but in lower-symmetry cases, the normal to (hkl) is not parallel to [hkl].

- Indices in *curly brackets* or *braces* such as {100} denote a family of planes and their normals. In cubic materials the symmetry makes them equivalent, just as the way angle brackets denote a family of directions. In non-cubic materials, <hkl> is not necessarily perpendicular to {hkl}.

Techniques

Some materials that have been analyzed crystallographically, such as proteins, do not occur naturally as crystals. Typically, such molecules are placed in solution and allowed to slowly crystallize through vapor diffusion. A drop of solution containing the molecule, buffer, and precipitants is sealed in a container with a reservoir containing a hygroscopic solution. Water in the drop diffuses to the reservoir, slowly increasing the concentration and allowing a crystal to form. If the concentration were to rise more quickly, the molecule would simply precipitate out of solution, resulting in disorderly granules rather than an orderly and hence usable crystal.

Once a crystal is obtained, data can be collected using a beam of radiation. Although many universities that engage in crystallographic research have their own X-ray producing equipment, synchrotrons are often used as X-ray sources, because of the purer and more complete patterns such sources can generate. Synchrotron sources also have a much higher intensity of X-ray beams, so data collection takes a fraction of the time normally necessary at weaker sources. Complementary neutron crystallography techniques are used to identify the positions of hydrogen atoms, since X-rays only interact very weakly with light elements such as hydrogen.

Producing an image from a diffraction pattern requires sophisticated mathematics and often an iterative process of modelling and refinement. In this process, the mathematically predicted diffraction patterns of an hypothesized or "model" structure are compared to the actual pattern generated by the crystalline sample. Ideally, researchers make several initial guesses, which through refinement all converge on the same answer. Models are refined until their predicted patterns match to as great a degree as can be achieved without radical revision of the model. This is a painstaking process, made much easier today by computers.

The mathematical methods for the analysis of diffraction data only apply to patterns, which in turn result only when waves diffract from orderly arrays. Hence crystallography applies for the most part only to crystals, or to molecules which can be coaxed to crystallize for the sake of measurement. In spite of this, a certain amount of molecular information can be deduced from patterns that are generated by fibers and powders, which while not as perfect as a solid crystal, may exhibit a degree of order. This level of order can be sufficient to deduce the structure of simple molecules, or to determine the coarse features of more complicated molecules. For example, the double-helical structure of DNA was deduced from an X-ray diffraction pattern that had been generated by a fibrous sample.

Crystal Growth

Crystals are used in semiconductor physics, engineering, as electro-optic devices etc., so there is an increasing demand for crystal. For years, Natural specimens were the only source of large, well formed crystals. The growth of crystals generally occurs by means of following sequence of process:

- Diffusion of the molecules of the crystallizing substance through the surrounding environment.

- Diffusion of these molecules over the surface of the crystal to special sites on the surface.

Today almost all naturally occurring crystals of interest have been synthesized successfully in the laboratory. It is now possible only by crystal growth techniques.

The growth aspect differs from crystal depending as their physics and chemical properties such as solubility, melting point, decomposition, phase change, etc.

Basics of Crystal Growth

To grow a crystal, the basic condition to be attained is the state of super saturation, followed by the process of nucleation. The information of super saturation and nucleation forms the basis of crystal growth.

Conditions for Growing Crystal

The growth of crystals from liquid and gaseous solutions, pure liquids and pure gases can only occur if some degree of super saturation or super cooling has been first achieved in the system. The attainment of the supersaturated state is essential for any crystallization operation and the degree of super saturation or deviation from the equilibrium saturated condition is the prime factor controlling the deposition process. Growth of crystals can be considered to compress these steps.

- Achievement of super saturation or super cooling.

- Formation of crystal nucleus of microscopic size.

- Successive growth of crystals to yield distinct faces.

Crystal Growth Techniques

Crystal growth is a challenging task and the technique followed for crystal growth depends upon the characteristics of the materials under investigation, such as its melting point, Volatile nature, solubility in water or other organic solvents and so on.

The basic growth methods available for crystal growth are broadly classified.

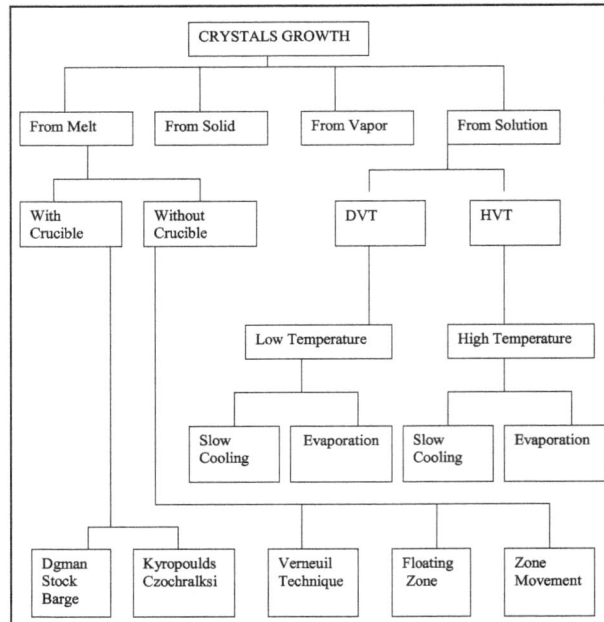

Classification of Crystal growth methods.

Growth from the Melt

Melt growth is the process of crystallization of fusion and resolidification of the pure material, crystallization from a melt on cooling the liquid below its freezing point. In this technique apart from possible contamination from crucible materials and surrounding atmosphere, no impurities are introduced in the growth process and the rate or growth is normally much higher than possible by other methods. Melt growth is commercially the most important method of crystal growth. The growth from melt can further be sub-grouped into various techniques:

- Bridgmann method.
- Czochralski method.
- Vernuil method.
- Zone melting method.
- Kyropoulos technique.
- Skull melting.

Bridgmann Method

This technique was named after its inventor Bridgemann in 1925, Stockbarger in 1938.

The Bridgmann technique is a method of growing single crystal ingots or boules. The method involves heating polycrystalline material in a container above its melting point and slowly cooling it from one end where a seed crystal is located. Single crystal material is progressively formed along the length of the container, the process can be carried out in a horizontal or vertical geometry.

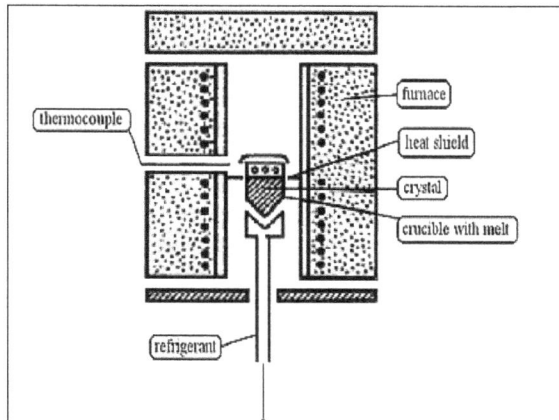

Bridgmann method.

Advantage

- This method is technically simple.

- This technique is low cost.

- Selecting the appropriate container can produce crystal of pre assigned diameter.

Disadvantage

- The compression of the solid by the contracting container during cooling can lead to the development of stresses high enough to nucleate dislocations in the material.

Czochralski Method (or) Pulling Technique

This method is widely used for growing semi conducting material crystal. The shape of the crystal is free from the constraint due to the shape of the crucible. In this method the charge is melted and maintained at a temperature slightly above the melting point. The pulling rod is lowered to just touch the melt. Since the rod is at lower temperature of melt occurs at the point tip of the pulling rod. The crystal is pulled slowly. The rate of pulling upon various factors like thermal conductivity, latent heat of fusion of charge and rate of cooling of the pulling rod. The seed is rotated to keep the grow crystal uniform and cylindrical.

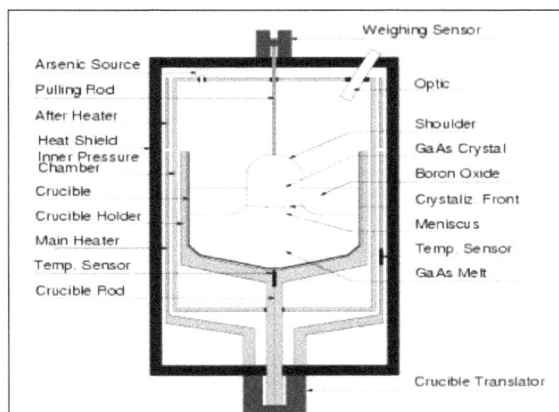

Czochralski method.

Advantage

- This method is used to grow large single crystals. Thus it is used extensively in the semi-conductor industry.

- There is no direct contact between the crucible walls and the crystal which helps to produce unstressed single crystal.

Disadvantage

- In general this method is not suitable for incongruently melting compounds and of course the need for a seed crystal of the same composition limits is used as tool for exploratory synthetic research.

Vernuil Method

The basis of vernuils method is as follows. Chemically pure fine powder which emerges through an Oxygen-hydrogen flame and falls onto the fused end of an oriented single crystal seed fixed to a lowering mechanism. The powder charge is fed from bankeei by mean of a special tapping mechanism. Coordinating the consumption of the charge, hydrogen and oxygen with the rate of descent of the seed ensures crystallization at a prescribed level of the apparatus.

Advantage

- There is no container which eliminates the problem of physical-chemical interaction between the melt and the container material.

- It is technically simple and the growth of crystal can be observed.

- Single crystal of ruby, sapphire etc., can be grown by this method. Single crystal in various Shapes like plates, disc, hemi-sphere and cones can be grown by this method.

Zone Melting Method

In this technique a liquid zone is created by melting a small amount of materials in a relatively large or long solid charge. Zone melting techniques basically enable one to manipulate distribution of soluble impurities or phases through a solid.

Advantage

- Zone melting technique is that impurities tend to be concentrated in the melted portion of the sample.

- The process sweeps them out of the sample and concentrates them at the end of the crystal bowl, which is then cut off and discarded.

- Thus this method is sometimes used to purify semiconductor crystals.

Kyropoulos Technique

In this technique, a cooled seed to initiate single crystal growth within the melt containing crucible. Heat removal continues by controlling the furnace temperature to grow the crystal.

Advantage

- The crystal is grown in a larger diameter.
- With the large diameter crystal we can make prisms, lenses and other optical components.

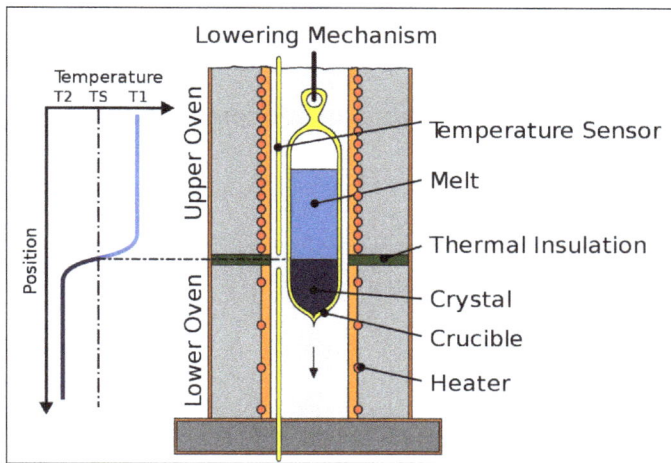

Kyropoulos Technique.

Skull Melting Process

Skull melting process is used for the growth of high melting point materials. Cubic zirconium is made using a radio-frequency "Skull crucible" system, a specialized melt process. This is a type of super-hot melt process used to produce the most widely accepted diamond imitation: Cubic zirconia(cz). Zirconium oxide, the main component in cubic zirconias has an extremely high melting point (4980 °F). Conventional (low) melt crucibles cannot be used. Therefore a crucible free technique is inevitable. The Skull melting process is used to produce Zirconia up to 10cm long.

Vapour Growth

Crystallization from Vapour is widely adopted to grow bulk crystal, epitaxial films, and thin coatings. Techniques for growing crystals from vapour is divided into two types they are:

- Chemical transport method.
- Physical transport method.

Chemical Transport Method

This method involves a chemical transport in which material is transported as a chemical compound (halide), which decomposes in the growth area. In this case depending on the nature of the reaction involved. The growth region may be either hotter or cooler than the source.

Physical Transport Method

This method involves in the direct transport of materials by evaporation or sublimation from a hot source zone to a cool region II-VI compounds (Zns, Cds) are widely grown by this method either in vaccum or with a moving gas stream. In both cases the growth can be suitable with seed crystals, which can either be of the material being grown or some other material with similar lattice spacing. In this case the substance evaporates and diffuses from hot end to a cooler growth end. In then, deposits in the form of single crystals.

Advantages

- Films can obtain by the close spaced transport method and decomposition of compounds.
- Crystal of silicon, diamond, gas, semiconductor compounds can be grown by this method.

Solution Growth

In this method, Crystals are grown from aqueous solution. This method is also widely practiced for producing bulk crystals. The four major types are:

- Low temperature solution growth.
- High temperature solution growth.
- Hydro Thermal growth.
- Gel Growth.

Low Temperature Solution Growth

This is a widely practiced method; the techniques used here are,

- Slow cooling method.
- Solvent Evaporation method.
- Temperature Gradient method

The solvent used here are water, ethyl alcohol, acetone, etc.

- Slow cooling method

  A saturated solution above the room temperature is poured in a crystallizer and thermally sealed. A seed crystal is suspended in the solution and the crystallizer is kept in a water thermostat, whose temperature is reducing according to a pre assigned plan, which results in the formation of large single crystals. The need to use a range of temperature is the origin of disadvantages. The possible range is usually small so that much of the solute remains in the solution at the end of run. To compensate for this effect, large volumes of solution are needed.

- Solvent Evaporation method

  In this method, an excess of a given solute is established by utilizing the difference between the rates of evaporation of the solvent and solute. In contrast to the cooling method, in which the total mass of the system remains constant, in the solvent evaporation method, the solution loses particles, which are weakly bound to other components, and therefore the volume of the solution decreases. In almost all cases, the vapour pressure of the solvent above the solution is higher than the vapour pressure of the solute and, therefore the solvent evaporates more rapidly and the solution becomes supersaturated. Usually, it is sufficient to allow the vapour formed above the solution to escape freely into the atmosphere. This is the oldest method of crystal growth and technically, it is very simple. Typical growth conditions involve temperature stabilization to about ±0.0005 °C and rates of evaporation of a few mm³/hr.

- Temperature Gradient method

The transport of material forms a hot region containing the source of the material to be grown, to a cooler region where solution is super saturated result in the crystal growth. A smaller variation in the temperature between the source and the crystal has larger effects on growth rate.

High Temperature Solution Growth

- The solvents are considered generally effective at temperatures above room temperature. Also the concepts of low temperature solution growth are applicable equally well. In the growth of crystals from high-temperature solutions, the constituents of the material to be crystallized are dissolved in a suitable solvent and crystallization occurs as the solution becomes critically supersaturated. The most widely used high temperature solution growth technique is the flux growth.

Hydro Thermal Growth

This is regarded as an intermediate case between growths from the vapour and solution. Growth occurs from aqueous solution at high temperature and pressure. The liquids from which the process starts are usually alkaline aqueous solutions. Temperatures are typically in the range 400-600 °C and the pressure involved is large (100-1000 of atmospheres). Growth is usually carried out in steel auto claves with gold or silver linings. The concentration gradient required to produce growth is provided by temperature difference (usually 10-100 °C) between the nutrient and growth areas. Those materials like calcite, alumina, antimony, etc., can be grown by this technique.

Advantages

- It occurs in air at a temperature much lower than the melting of the crystallizing substance.
- Single crystals of diamond, barium titanates are to be grown by using this method.

Disadvantages

- At elevated temperature of crystal growth there are some disadvantages.

Gel Growth Method

The principle used in this crystal growth techniques is very simple solution of two suitable compounds give rise to required crystalline substance by mere chemical reactions, crystallization takes place according to chemical reaction equation as shown below:

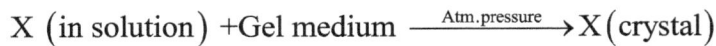

$$X \text{ (in solution)} + \text{Gel medium} \xrightarrow{\text{Atm.pressure}} X \text{ (crystal)}$$

Various Types of Gels

- Physical gel: Gel which is obtained by Physical process such as cooling is called physical gel. Eg. Gelatin, clay.

- Chemical gel: Gels formed by chemical reaction such as hydrolysis (or) polymerization are called chemical gels. E.g. Silica, Polyacryalamide.

Advantages

- It prevents turbulence and formation of good crystals by providing a framework of nucleation.

- The convection is absent in growth experiments.

- The high degree of perfection and lesser number of defects have been observed in gel growth.

- The Gel method has also been applied to the study of crystal formation in human system such as cholesterol stores, Sex hormones.

MATERIALS INFORMATICS

Materials informatics is a field of study that applies the principles of informatics to materials science and engineering to better understand the use, selection, development, and discovery of materials. This is an emerging field, with a goal to achieve high-speed and robust acquisition, management, analysis, and dissemination of diverse materials data with the goal of greatly reducing the time and risk required to develop, produce, and deploy a new materials, which generally takes longer than 20 years.

This field of endeavor is not limited to some traditional understandings of the relationship between materials and information. Some more narrow interpretations include combinatorial chemistry, Process Modeling, materials property databases, materials data management and product life cycle management. Materials informatics is at the convergence of these concepts, but also transcends them and has the potential to achieve greater insights and deeper understanding by applying lessons learned from data gathered on one type of material to others. By gathering appropriate meta data, the value of each individual data point can be greatly expanded.

The concept of materials informatics is addressed by the Materials Research Society. For example, materials informatics was the theme of the December 2006 issue of the MRS Bulletin. The issue was guest-edited by John Rodgers of Innovative Materials, Inc., and David Cebon of Cambridge University, who described the "high payoff for developing methodologies that will accelerate the insertion of materials, thereby saving millions of investment dollars".

The editors focused on the limited definition of materials informatics as primarily focused on computational methods to process and interpret data. They stated that "specialized informatics tools for data capture, management, analysis, and dissemination" and "advances in computing power, coupled with computational modeling and simulation and materials properties databases" will enable such accelerated insertion of materials.

A broader definition of materials informatics goes beyond the use of computational methods to carry out the same experimentation, viewing materials informatics as a framework in which a measurement or computation is one step in an information-based learning process that uses the power of a collective to achieve greater efficiency in exploration. When properly organized, this framework crosses materials boundaries to uncover fundamental knowledge of the basis of physical, mechanical, and engineering properties.

Challenges

While there are many that believe in the future of informatics in the materials development and scaling process, many challenges remain. Hill, et. al., write that "Today, the materials community faces serious challenges to bringing about this data-accelerated research paradigm, including diversity of research areas within materials, lack of data standards, and missing incentives for sharing, among others. Nonetheless, the landscape is rapidly changing in ways that should benefit the entire materials research enterprise". This remaining tension between traditional materials development methodologies and the use of more computationally, machine learning, and analytics approaches will likely exist for some time as the materials industry overcomes some of the cultural barriers necessary to fully embrace such new ways of thinking.

POLYMER ENGINEERING

Polymer engineering is generally an engineering field that designs, analyses, and modifies polymer materials. Polymer engineering covers aspects of the petrochemical industry, polymerization, structure and characterization of polymers, properties of polymers, compounding and processing of polymers and description of major polymers, structure property relations and applications.

Classification

The basic division of polymers into thermoplastics, elastomers and thermosets helps define their areas of application.

Thermoplastics

Thermoplastic refers to a plastic that has heat softening and cooling hardening properties. Most of the plastics we use in our daily lives fall into this category. It becomes soft and even flows when heated, and the cooling becomes hard. This process is reversible and can be repeated. Thermoplastics have relatively low tensile moduli, but also have lower densities and properties such as transparency which make them ideal for consumer products and medical products. They include polyethylene, polypropylene, nylon, acetal resin, polycarbonate and PET, all of which are widely used materials.

Elastomers

An elastomer generally refers to a material that can be restored to its original state after removal of an external force, whereas a material having elasticity is not necessarily an elastomer. The elastomer is only deformed under weak stress, and the stress can be quickly restored to a polymer material close to the original state and size. Elastomers are polymers which have very low moduli and show reversible extension when strained, a valuable property for vibration absorption and damping. They may either be thermoplastic (in which case they are known as Thermoplastic elastomers) or crosslinked, as in most conventional rubber products such as tyres. Typical rubbers used conventionally include natural rubber, nitrile rubber, polychloroprene, polybutadiene, styrene-butadiene and fluorinated rubbers.

Thermosets

A thermosetting resin is used as a main component, and a plastic which forms a product is formed by a cross-linking curing process in combination with various necessary additives. It is liquid in the early stage of the manufacturing or molding process, and it is insoluble and infusible after curing, and it cannot be melted or softened again. Common thermosetting plastics are phenolic plastics, epoxy plastics, aminoplasts, unsaturated polyesters, alkyd plastics, and the like. Thermoset plastics and thermoplastics together constitute the two major components of synthetic plastics. Thermosetting plastics are divided into two types: formaldehyde cross-linking type and other cross-linking type.

Thermosets includes phenolic resins, polyesters and epoxy resins, all of which are used widely in composite materials when reinforced with stiff fibers such as fiberglass and aramids. Since cross-linking stabilises the thermoset polymer matrix of these materials, they have physical properties more similar to traditional engineering materials like steel. However, their very much lower densities compared with metals makes them ideal for lightweight structures. In addition, they suffer less from fatigue, so are ideal for safety-critical parts which are stressed regularly in service.

Materials

Plastic

Plastic is a polymer compound which is polymerized by polyaddition polymerization and polycondensation. It is free to change the composition and shape. It is made up of synthetic resins and fillers, plasticizers, stabilizers, lubricants, colorants and other additives. The main component of plastic is resin. Resin means that the polymer compound has not been added with various additives. The term resin was originally named for the secretion of oil from plants and animals, such as rosin and shellac. Resin accounts for approximately 40% - 100% of the total weight of the plastic. The basic properties of plastics are mainly determined by the nature of the resin, but additives also play an important role. Some plastics are basically made of synthetic resins, with or without additives such as plexiglass, polystyrene, etc.

Fiber

Fiber refers to a continuous or discontinuous filament of one substance. Animals and plant fibers play an important role in maintaining tissue. Fibers are widely used and can be woven into good threads, thread ends and hemp ropes. They can also be woven into fibrous layers when making paper or feel. They are also commonly used to make other materials together with other materials to form composites. Therefore, whether it is natural or synthetic fiber filamentous material. In modern life, the application of fiber is ubiquitous, and there are many high-tech products.

Rubber

Rubber refers to highly elastic polymer materials and reversible shapes. It is elastic at room temperature and can be deformed with a small external force. After removing the external force, it can return to the original state. Rubber is a completely amorphous polymer with a low glass transition temperature and a large molecular weight, often greater than several hundred thousand. Highly elastic polymer compounds can be classified into natural rubber and synthetic rubber. Natural rubber processing extracts gum rubber and grass rubber from plants; synthetic rubber is polymerized by various monomers. Rubber can be used as elastic, insulating, water-impermeable air-resistant materials.

Applications

B-2 Spirit stealth bomber.

Polyethylene

Commonly used polyethylenes can be classified into low density polyethylene (LDPE), high density polyethylene (HDPE), and linear low density polyethylene (LLDPE). Among them, HDPE has better thermal, electrical and mechanical properties, while LDPE and LLDPE have better flexibility, impact properties and film forming properties. LDPE and LLDPE are mainly used for plastic bags, plastic wraps, bottles, pipes and containers; HDPE is widely used in various fields such as film, pipelines and daily necessities because its resistance to many different solvents.

Polypropylene

Polypropylene is widely used in various applications due to its good chemical resistance and weldability. It has lowest density among commodity plastics. It is commonly used in packaging applications, consumer goods, automatic applications and medical applications. Polypropylene sheets are widely used in industrial sector to produce acid and chemical tanks, sheets, pipes, Returnable

Transport Packaging (RTP), etc. because of its properties like high tensile strength, resistance to high temperatures and corrosion resistance.

Composites

A time-trial carbon fibre composite bicycle with
aerodynamic wheels and aero bars.

Typical uses of composites are monocoque structures for aerospace and automobiles, as well as more mundane products like fishing rods and bicycles. The stealth bomber was the first all-composite aircraft, but many passenger aircraft like the Airbus and the Boeing 787 use an increasing proportion of composites in their fuselages, such as hydrophobic melamine foam. The quite different physical properties of composites gives designers a much greater freedom in shaping parts, which is why composite products often look different to conventional products. On the other hand, some products such as drive shafts, helicopter rotor blades, and propellers look identical to metal precursors owing to the basic functional needs of such components.

Biomedical Applications

Biodegradable polymers are widely used materials for many biomedical and pharmaceutical applications. They are considered vey promising for controlled drug delivery devices. Biodegradable polymers also offer great potential for wound management, orthopaedic devices, dental applications and tissue engineering. Not like non biodegradable polymers, they won't require a second step of a removal from body. Biodegradable polymers will break down and are absorbed by the body after they served their purpose. Since 1960, polymers prepared from glycolic acid and lactic acid have found a multitude of uses in the medical industry. Polylactates (PLAs) are popular for drug delivery system due to their fast and adjustable degradation rate.

Membrane Technologies

Membrane techniques are successfully used in the separation in the liquid and gas systems for years, and the polymeric membranes are used most commonly because they have lower cost to produce and are easy to modify their surface, which make them suitable in different separation processes. Polymers helps in many fields including the application for separation of biological active compounds, proton exchange membranes for fuel cells and membrane contractors for carbon dioxide capture process.

POLYMER SCIENCE

Polymer science or macromolecular science is a subfield of materials science concerned with polymers, primarily synthetic polymers such as plastics and elastomers. The field of polymer science includes researchers in multiple disciplines including chemistry, physics, and engineering.

The first modern example of polymer science is Henri Braconnot's work in the 1830s. Henri, along with Christian Schönbein and others, developed derivatives of the natural polymer cellulose, producing new, semi-synthetic materials, such as celluloid and cellulose acetate. The term "polymer" was coined in 1833 by Jöns Jakob Berzelius, though Berzelius did little that would be considered polymer science in the modern sense. In the 1840s, Friedrich Ludersdorf and Nathaniel Hayward independently discovered that adding sulfur to raw natural rubber (polyisoprene) helped prevent the material from becoming sticky. In 1844 Charles Goodyear received a U.S. patent for vulcanizing natural rubber with sulfur and heat. Thomas Hancock had received a patent for the same process in the UK the year before. This process strengthened natural rubber and prevented it from melting with heat without losing flexibility. This made practical products such as waterproofed articles possible. It also facilitated practical manufacture of such rubberized materials. Vulcanized rubber represents the first commercially successful product of polymer research. In 1884 Hilaire de Chardonnet started the first artificial fiber plant based on regenerated cellulose, or viscose rayon, as a substitute for silk, but it was very flammable. In 1907 Leo Baekeland invented the first synthetic plastic, a thermosetting phenol–formaldehyde resin called Bakelite.

Despite significant advances in polymer synthesis, the molecular nature of polymers was not understood until the work of Hermann Staudinger in 1922. Prior to Staudinger's work, polymers were understood in terms of the association theory or aggregate theory, which originated with Thomas Graham in 1861. Graham proposed that cellulose and other polymers were colloids, aggregates of molecules having small molecular mass connected by an unknown intermolecular force. Hermann Staudinger was the first to propose that polymers consisted of long chains of atoms held together by covalent bonds. It took over a decade for Staudinger's work to gain wide acceptance in the scientific community, work for which he was awarded the Nobel Prize in 1953.

The World War II era marked the emergence of a strong commercial polymer industry. The limited or restricted supply of natural materials such as silk and rubber necessitated the increased production of synthetic substitutes, such as nylon and synthetic rubber. In the intervening years, the development of advanced polymers such as Kevlar and Teflon have continued to fuel a strong and growing polymer industry.

The growth in industrial applications was mirrored by the establishment of strong academic programs and research institute. In 1946, Herman Mark established the Polymer Research Institute at Brooklyn Polytechnic, the first research facility in the United States dedicated to polymer research. Mark is also recognized as a pioneer in establishing curriculum and pedagogy for the field of polymer science. In 1950, the POLY division of the American Chemical Society was formed, and has since grown to the second-largest division in this association with nearly 8,000 members. Fred W. Billmeyer, Jr., a Professor of Analytical Chemistry had once said that "although the scarcity of education in polymer science is slowly diminishing but it is still evident in many areas. What is most unfortunate is that it appears to exist, not because of a lack of awareness but, rather, a lack of interest."

METALLURGY

Metallurgy is the art and science of extracting metals from their ores and modifying the metals for use. Metallurgy customarily refers to commercial as opposed to laboratory methods. It also concerns the chemical, physical, and atomic properties and structures of metals and the principles whereby metals are combined to form alloys.

Earliest Development

Gold can be agglomerated into larger pieces by cold hammering, but native copper cannot, and an essential step toward the Metal Age was the discovery that metals such as copper could be fashioned into shapes by melting and casting in molds; among the earliest known products of this type are copper axes cast in the Balkans in the 4th millennium BCE. Another step was the discovery that metals could be recovered from metal-bearing minerals. These had been collected and could be distinguished on the basis of colour, texture, weight, and flame colour and smell when heated. The notably greater yield obtained by heating native copper with associated oxide minerals may have led to the smelting process, since these oxides are easily reduced to metal in a charcoal bed at temperatures in excess of 700 °C (1,300 °F), as the reductant, carbon monoxide, becomes increasingly stable. In order to effect the agglomeration and separation of melted or smelted copper from its associated minerals, it was necessary to introduce iron oxide as a flux. This further step forward can be attributed to the presence of iron oxide gossan minerals in the weathered upper zones of copper sulfide deposits.

Bronze

In many regions, copper-arsenic alloys, of superior properties to copper in both cast and wrought form, were produced in the next period. This may have been accidental at first, owing to the similarity in colour and flame colour between the bright green copper carbonate mineral malachite and the weathered products of such copper-arsenic sulfide minerals as enargite, and it may have been followed later by the purposeful selection of arseniccompounds based on their garlic odour when heated.

Arsenic contents varied from 1 to 7 percent, with up to 3 percent tin. Essentially arsenic-free copper alloys with higher tin content—in other words, true bronze—seem to have appeared between 3000 and 2500 BCE, beginning in the Tigris-Euphrates delta. The discovery of the value of tin may have occurred through the use of stannite, a mixed sulfide of copper, iron, and tin, although this mineral is not as widely available as the principal tin mineral, cassiterite, which must have been the eventual source of the metal. Cassiterite is strikingly dense and occurs as pebbles in alluvial deposits together with arsenopyrite and gold; it also occurs to a degree in the iron oxide gossans.

While there may have been some independent development of bronze in varying localities, it is most likely that the bronze culture spread through trade and the migration of peoples from the Middle East to Egypt, Europe, and possibly China. In many civilizations the production of copper, arsenical copper, and tin bronze continued together for some time. The eventual disappearance of copper-arsenic alloys is difficult to explain. Production may have been based on minerals that were not widely available and became scarce, but the relative scarcity of tin minerals did not prevent a substantial trade in that metal over considerable distances. It may be that tin bronzes were eventually preferred owing to the chance of contracting arsenic poisoning from fumes produced by the oxidation of arsenic-containing minerals.

As the weathered copper ores in given localities were worked out, the harder sulfide ores beneath were mined and smelted. The minerals involved, such as chalcopyrite, a copper-iron sulfide, needed an oxidizing roast to remove sulfur as sulfur dioxide and yield copper oxide. This not only required greater metallurgical skill but also oxidized the intimately associated iron, which, combined with the use of iron oxide fluxes and the stronger reducing conditions produced by improved smelting furnaces, led to higher iron contents in the bronze.

Iron

It is not possible to mark a sharp division between the Bronze Age and the Iron Age. Small pieces of iron would have been produced in copper smelting furnaces as iron oxide fluxes and iron-bearing copper sulfide ores were used. In addition, higher furnace temperatures would have created more strongly reducing conditions (that is to say, a higher carbon monoxide content in the furnace gases). An early piece of iron from a trackway in the province of Drenthe, Netherlands, has been dated to 1350 BCE, a date normally taken as the Middle Bronze Age for this area. In Anatolia, on the other hand, iron was in use as early as 2000 BCE. There are also occasional references to iron in even earlier periods, but this material was of meteoric origin.

Once a relationship had been established between the new metal found in copper smelts and the ore added as flux, the operation of furnaces for the production of iron alone naturally followed. Certainly, by 1400 BCE in Anatolia, iron was assuming considerable importance, and by 1200–1000 BCE it was being fashioned on quite a large scale into weapons, initially dagger blades. For this reason, 1200 BCE has been taken as the beginning of the Iron Age. Evidence from excavations indicates that the art of iron making originated in the mountainous country to the south of the Black Sea, an area dominated by the Hittites. Later the art apparently spread to the Philistines, for crude furnaces dating from 1200 BCE have been unearthed at Gerar, together with a number of iron objects.

Smelting of iron oxide with charcoal demanded a high temperature, and, since the melting temperature of iron at 1,540 °C (2,800 °F) was not attainable then, the product was merely a spongy mass of pasty globules of metal intermingled with a semiliquid slag. This product, later known as bloom, was hardly usable as it stood, but repeated reheating and hot hammering eliminated much of the slag, creating wrought iron, a much better product.

The properties of iron are much affected by the presence of small amounts of carbon, with large increases in strength associated with contents of less than 0.5 percent. At the temperatures then attainable—about 1,200 °C (2,200 °F)—reduction by charcoal produced an almost pure iron, which was soft and of limited use for weapons and tools, but when the ratio of fuel to ore was increased and furnace drafting improved with the invention of better bellows, more carbon was absorbed by the iron. This resulted in blooms and iron products with a range of carbon contents, making it difficult to determine the period in which iron may have been purposely strengthened by carburizing, or reheating the metal in contact with excess charcoal.

Carbon-containing iron had the further great advantage that, unlike bronze and carbon-free iron, it could be made still harder by quenching—i.e., rapid cooling by immersion in water. There is no evidence for the use of this hardening process during the early Iron Age, so that it must have been either unknown then or not considered advantageous, in that quenchingrenders iron very brittle and has to be followed by tempering, or reheating at a lower temperature, to restore toughness. What seems to have been established early on was a practice of repeated cold forging and annealing at 600–700 °C (1,100–1,300 °F), a temperature naturally achieved in a simple fire. This practice is common in parts of Africa even today.

By 1000 BCE iron was beginning to be known in central Europe. Its use spread slowly westward. Iron making was fairly widespread in Great Britain at the time of the Roman invasion in 55 BCE. In Asia iron was also known in ancient times, in China by about 700 BCE.

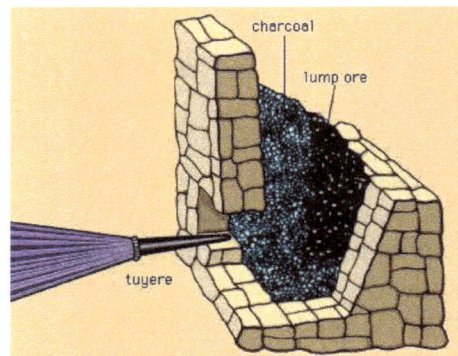

Catalan hearth or forge used for smelting iron ore until relatively recent times.
The method of charging fuel and ore and the approximate position of the
nozzle supplied with air by a bellows are shown.

If the fuel-to-ore ratio in such a furnace was kept high, and if the furnace reached temperatures sufficiently hot for substantial amounts of carbon to be absorbed into the iron, then the melting point of the metal would be lowered and the bloom would melt. This would dissolve even more carbon, producing a liquid cast iron of up to 4 percent carbon and with a relatively low melting temperature of 1,150 °C (2,100 °F). The cast iron would collect in the base of the furnace, which technically would be a blast furnace rather than a bloomery in that the iron would be withdrawn as a liquid rather than a solid lump.

While the Iron Age peoples of Anatolia and Europe on occasion may have accidently made cast iron, which is chemically the same as blast-furnace iron, the Chinese were the first to realize its advantages. Although brittle and lacking the strength, toughness, and workability of steel, it was useful for making cast bowls and other vessels. In fact, the Chinese, whose Iron Age began about 500 BCE, appear to have learned to oxidize the carbon from cast iron in order to produce steel or wrought iron indirectly, rather than through the direct method of starting from low-carbon iron.

After 1500

During the 16th century, metallurgical knowledge was recorded and made available. Two books were especially influential. One, by the Italian Vannoccio Biringuccio, was entitled De la pirotechnia (Eng. trans., The Pirotechnia of Vannoccio Biringuccio, 1943). The other, by the German Georgius Agricola, was entitled De re metallica. Biringuccio was essentially a metalworker, and his book dealt with smelting, refining, and assay methods (methods for determining the metal content of ores) and covered metal casting, molding, core making, and the production of such commodities as cannons and cast-iron cannonballs. His was the first methodical description of foundry practice.

Agricola, on the other hand, was a miner and an extractive metallurgist; his book considered prospecting and surveying in addition to smelting, refining, and assay methods. He also described the processes used for crushing and concentrating the ore and then, in some detail, the methods of assaying to determine whether ores were worth mining and extracting. Some of the metallurgical practices he described are retained in principle today.

Ferrous Metals

From 1500 to the 20th century, metallurgical development was still largely concerned with improved technology in the manufacture of iron and steel. In England, the gradual exhaustion of timber led first to prohibitions on cutting of wood for charcoal and eventually to the introduction of coke, derived from coal, as a more efficient fuel. Thereafter, the iron industry expanded rapidly in Great Britain, which became the greatest iron producer in the world. The crucible process for making steel, introduced in England in 1740, by which bar iron and added materials were placed in clay cruciblesheated by coke fires, resulted in the first reliable steel made by a melting process.

One difficulty with the bloomery process for the production of soft bar iron was that, unless the temperature was kept low (and the output therefore small), it was difficult to keep the carbon content low enough so that the metal remained ductile. This difficulty was overcome by melting high-carbon pig iron from the blast furnace in the puddling process, invented in Great Britain in 1784. In it, melting was accomplished by drawing hot gases over a charge of pig iron and iron ore held on the furnace hearth. During its manufacture the product was stirred with iron rabbles (rakes), and, as it became pasty with loss of carbon, it was worked into balls, which were subsequently forged or rolled to a useful shape. The product, which came to be known as wrought iron, was low in elements that contributed to the brittleness of pig iron and contained enmeshed slag particles that became elongated fibres when the metal was forged. Later, the use of a rolling mill equipped with grooved rolls to make wrought-iron bars was introduced.

The most important development of the 19th century was the large-scale production of cheap steel. Prior to about 1850, the production of wrought iron by puddling and of steel by crucible melting had been conducted in small-scale units without significant mechanization. The first change was the development of the open-hearth furnace by William and Friedrich Siemens in Britain and by Pierre and Émile Martin in France. Employing the regenerative principle, in which outgoing combusted gases are used to heat the next cycle of fuel gas and air, this enabled high temperatures to be achieved while saving on fuel. Pig iron could then be taken through to molten iron or low-carbon steel without solidification, scrap could be added and melted, and iron ore could be melted into the slag above the metal to give a relatively rapid oxidation of carbon and silicon—all on a much enlarged scale. Another major advance was Henry Bessemer's process, patented in 1855 and first operated in 1856, in which air was blown through molten pig iron from tuyeres set into the bottom of a pear-shaped vessel called a converter. Heat released by the oxidation of dissolved silicon, manganese, and carbon was enough to raise the temperature above the melting point of the refined metal (which rose as the carbon content was lowered) and thereby maintain it in the liquid state. Very soon Bessemer had tilting converters producing 5 tons in a heat of one hour, compared with four to six hours for 50 kilograms (110 pounds) of crucible steel and two hours for 250 kilograms of puddled iron.

Neither the open-hearth furnace nor the Bessemer converter could remove phosphorus from the metal, so that low-phosphorus raw materials had to be used. This restricted their use from areas where phosphoric ores, such as those of the Minette range in Lorraine, were a main European source of iron. The problem was solved by Sidney Gilchrist Thomas, who demonstrated in 1876 that a basic furnace lining consisting of calcined dolomite, instead of an acidic lining of siliceous materials, made it possible to use a high-lime slag to dissolve the phosphates formed by the oxidation of phosphorus in the pig iron. This principle was eventually applied to both open-hearth furnaces and Bessemer converters.

As steel was now available at a fraction of its former cost, it saw an enormously increased use for engineering and construction. Soon after the end of the century it replaced wrought iron in virtually every field. Then, with the availability of electric power, electric-arc furnaces were introduced for making special and high-alloy steels. The next significant stage was the introduction of cheap oxygen, made possible by the invention of the Linde-Frankel cycle for the liquefaction and fractional distillation of air. The Linz-Donawitz process, invented in Austria shortly after World War II, used oxygen supplied as a gas from a tonnage oxygen plant, blowing it at supersonic velocity into the top of the molten iron in a converter vessel. As the ultimate development of the Bessemer/Thomas process, oxygen blowing became universally employed in bulk steel production.

Light Metals

Another important development of the late 19th century was the separation from their ores, on a substantial scale, of aluminum and magnesium. In the earlier part of the century, several scientists had made small quantities of these light metals, but the most successful was Henri-Étienne Sainte-Claire Deville, who by 1855 had developed a method by which cryolite, a double fluoride of aluminum and sodium, was reduced by sodium metal to aluminum and sodium fluoride. The process was very expensive, but cost was greatly reduced when the American chemist Hamilton Young Castner developed an electrolytic cell for producing cheaper sodium in 1886. At the same time, however, Charles M. Hall in the United States and Paul-Louis-Toussaint Héroult in France announced their essentially identical processes for aluminum extraction, which were also based on electrolysis. Use of the Hall-Héroult process on an industrial scale depended on the replacement of storage batteries by rotary power generators; it remains essentially unchanged to this day.

Welding

One of the most significant changes in the technology of metals fabrication has been the introduction of fusion welding during the 20th century. Before this, the main joining processes were riveting and forge welding. Both had limitations of scale, although they could be used to erect substantial structures. In 1895 Henry-Louis Le Chatelier stated that the temperature in an oxyacetylene flame was 3,500 °C (6,300 °F), some 1,000 °C higher than the oxyhydrogen flame already in use on a small scale for brazing and welding. The first practical oxyacetylene torch, drawing acetylene from cylinders containing acetylene dissolved in acetone, was produced in 1901. With the availability of oxygen at even lower cost, oxygen cutting and oxyacetylene welding became established procedures for the fabrication of structural steel components.

The metal in a join can also be melted by an electric arc, and a process using a carbon as a negative electrode and the workpiece as a positive first became of commercial interest about 1902. Striking an arc from a coated metal electrode, which melts into the join, was introduced in 1910. Although it was not widely used until some 20 years later, in its various forms it is now responsible for the bulk of fusion welds.

Metallography

The 20th century has seen metallurgy change progressively, from an art or craft to a scientific discipline and then to part of the wider discipline of materials science. In extractive metallurgy, there has been the application of chemical thermodynamics, kinetics, and chemical engineering, which has enabled a better understanding, control, and improvement of existing processes and the generation of new ones. In physical metallurgy, the study of relationships between macrostructure, microstructure, and atomic structure on the one hand and physical and mechanical properties on the other has broadened from metals to other materials such as ceramics, polymers, and composites.

This greater scientific understanding has come largely from a continuous improvement in microscopic techniques for metallography, the examination of metal structure. The first true metallographer was Henry Clifton Sorby of Sheffield, England, who in the 1860s applied light microscopy to the polished surfaces of materials such as rocks and meteorites. Sorby eventually succeeded in making photomicrographic records, and by 1885 the value of metallography was appreciated throughout Europe, with particular attention being paid to the structure of steel. For example, there was eventual acceptance, based on micrographic evidence and confirmed by the introduction of X-ray diffraction by William Henry and William Lawrence Bragg in 1913, of the allotropy of iron and its relationship to the hardening of steel. During subsequent years there were advances in the atomic theory of solids; this led to the concept that, in nonplastic materials such as glass, fracture takes place by the propagationof preexisting cracklike defects and that, in metals, deformation takes place by the movement of dislocations, or defects in the atomic arrangement, through the crystalline matrix. Proof of these concepts came with the invention and development of the electron microscope; even more powerful field ion microscopes and high-resolution electron microscopes now make it possible to detect the position of individual atoms.

Another example of the development of physical metallurgy is a discovery that revolutionized the use of aluminum in the 20th century. Originally, most aluminum was used in cast alloys, but the discovery of age hardening by Alfred Wilm in Berlin about 1906 yielded a material that was twice as strong with only a small change in weight. In Wilm's process, a solute such as magnesium or copper is trapped in supersaturated solid solution, without being allowed to precipitate out, by quenching the aluminum from a higher temperature rather than slowly cooling it. The relatively soft aluminum alloy that results can be mechanically formed, but, when left at room temperature or heated at low temperatures, it hardens and strengthens. With copper as the solute, this type of material came to be known by the trade name Duralumin. The advances in metallography described above eventually provided the understanding that age hardening is caused by the dispersion of very fine precipitates from the supersaturated solid solution; this restricts the movement of the dislocations that are essential to crystal deformation and thus raises the strength of the metal. The principles of precipitation hardening have been applied to the strengthening of a large number of alloys.

Extractive Metallurgy

Following separation and concentration by mineral processing, metallic minerals are subjected to extractive metallurgy, in which their metallic elements are extracted from chemical compound form and refined of impurities.

Metallic compounds are frequently rather complex mixtures (those treated commercially are for the most part sulfides, oxides, carbonates, arsenides, or silicates), and they are not often types that permit extraction of the metalby simple, economical processes. Consequently, before extractive metallurgy can effect the separation of metallic elements from the other constituents of a compound, it must often convert the compound into a type that can be more readily treated. Common practice is to convert metallic sulfides to oxides, sulfates, or chlorides; oxides to sulfates or chlorides; and carbonates to oxides. The processes that accomplish all this can be categorized as either pyrometallurgy or hydrometallurgy. Pyrometallurgy involves heating operations such as roasting, in which compounds are converted at temperatures just below their melting points, and smelting, in which all the constituents of an ore or concentrate are completely melted and separated into two liquid layers, one containing the valuable metals and the other the waste rock. Hydrometallurgy consists of such operations as leaching, in which metallic compounds are selectively dissolved from an ore by an aqueous solvent, and electrowinning, in which metallic ions are deposited onto an electrode by an electric current passed through the solution.

Extraction is often followed by refining, in which the level of impurities is brought lower or controlled by pyrometallurgical, electrolytic, or chemical means. Pyrometallurgical refining usually consists of the oxidizing of impurities in a high-temperature liquid bath. Electrolysis is the dissolving of metal from one electrode of an electrolytic cell and its deposition in a purer form onto the other electrode. Chemical refining involves either the condensation of metal from a vapour or the selective precipitation of metal from an aqueous solution.

The processes to be used in extraction and refining are selected to fit into an overall pattern, with the product from the first process becoming the feed material of the second process, and so on. It is quite common for hydrometallurgical, pyrometallurgical, and electrolytic processes to be used one after another in the treatment of a single metal. The choices depend on several conditions. One of these is that certain types of metallic compounds lend themselves to easiest extraction by certain methods; for example, oxides and sulfates are readily dissolved in leach solutions, while sulfides are only slightly soluble. Another condition is the degree of purity, which can vary from one type of extraction to another. Zinc production illustrates this, in that zinc metal produced by pyrometallurgical retort or blast-furnace operations is 98 percent pure, with traces of lead, iron, and cadmium. This is adequate for galvanizing, but the preferred purity for die-casting (99.99 percent) must be obtained hydrometallurgically, from the electrolysis of a zinc sulfate solution. Also to be considered in choosing a processing method is the recovery of particular impurities that may have value themselves as by-products. One example is copper refining: the pyrometallurgical refining of blister copper removes many impurities, but it does not recover or remove silver or gold; the precious metals are recovered, however, by subsequent electrolytic refining.

Pyrometallurgy

Two of the most common pyrometallurgical processes, in both extraction and refining, are oxidation and reduction. In oxidation, metals having a great affinity for oxygen selectively combine with it to form metallic oxides; these can be treated further in order to obtain a pure metal or can be separated and discarded as a waste product. Reduction can be viewed as the reverse of oxidation. In this process, a metallic oxide compound is fed into a furnace along with a reducing agent such as carbon. The metal releases its combined oxygen, which recombines with the carbon to form a new carbonaceous oxide and leaves the metal in an uncombined form.

Oxidation and reduction reactions are either exothermic (energy-releasing) or endothermic (energy-absorbing). One example of an exothermic reaction is the oxidation of iron sulfide (FeS) to form iron oxide (FeO) and sulfur dioxide (SO_2) gas:

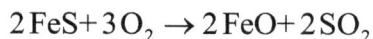

$$2\,FeS + 3\,O_2 \rightarrow 2\,FeO + 2\,SO_2$$

This process gives off large quantities of heat beyond that required to initiate the reaction. One endothermic reaction is the smelting reduction of zinc oxide (ZnO) by carbon monoxide (CO) to yield zinc (Zn) metal and carbon dioxide (CO_2):

$$ZnO + CO \rightarrow Zn + CO_2$$

For this reaction to proceed at a reasonable rate, external heat must be supplied to maintain the temperature at 1,300 to 1,350 °C (2,375 to 2,450 °F).

Roasting

As stated above, for those instances in which a metal-bearing compound is not in a chemical form that permits the metal to be easily and economically removed, it is necessary first to change it into some other compound. The preliminary treatment that is commonly used to do this is roasting.

Processes

There are several different types of roast, each one intended to produce a specific reaction and to yield a roasted product (or calcine) suitable for the particular processing operation to follow. The roasting procedures are:

1. Oxidizing roasts, which remove all or part of the sulfur from sulfide metal compounds, replacing the sulfides with oxides. (The sulfur removed goes off as sulfur dioxide gas.) Oxidizing roasts are exothermic.

2. Sulfatizing roasts, which convert certain metals from sulfides to sulfates. Sulfatizing roasts are exothermic.

3. Reducing roasts, which lower the oxide state or even completely reduce an oxide to a metal. Reducing roasts are exothermic.

4. Chloridizing roasts, or chlorination, which change metallic oxides to chlorides by heating with a chlorine source such as chlorine gas, hydrochloric acid gas, ammonium chloride, or sodium chloride. These reactions are exothermic.

5. Volatilizing roasts, which eliminate easily volatilized oxides by converting them to gases.

6. Calcination, in which solid material is heated to drive off either carbon dioxide or chemically combined water. Calcination is an endothermic reaction.

Roasters

Each of the above processes can be carried out in specialized roasters. The types most commonly in use are fluidized-bed, multiple-hearth, flash, chlorinator, rotary kiln, and sintering machine (or blast roaster).

Fluidized-bed roasters have found wide acceptance because of their high capacity and efficiency. They can be used for oxidizing, sulfatizing, and volatilizing roasts. The roaster is a refractory-lined, upright cylindrical steel shell with a grate bottom through which air is blown in sufficient volume to keep fine, solid feed particles in suspension and give excellent gas-solid contact. The ore feed can be introduced dry or as a water suspension through a downpipe into the turbulent layer zone of the roaster. Discharge of the roasted calcines is through a side overflow pipe.

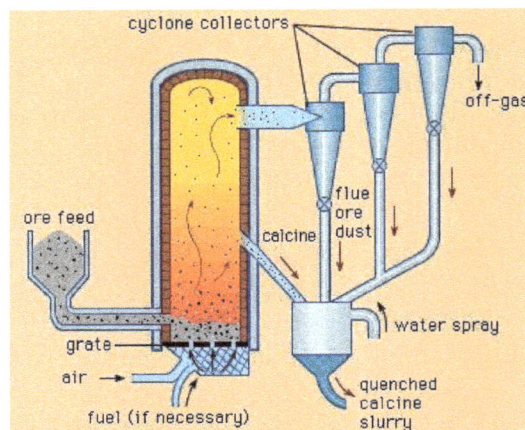

Schematic diagram of a fluidized-bed roaster.

Multiple-hearth roasters also have found wide acceptance in that they can be used for oxidizing, sulfatizing, chloridizing, volatilizing, reducing, and calcining processes. The roaster is a refractory-lined, vertical cylindrical steel shell in which are placed a number of superimposed refractory hearths. A slowly rotating central shaft turns rabble arms on each hearth both to stir the roasting material and to push it into drop holes leading to the hearth below. Feed material is fed to the top hearth, and, as it follows a zig-zag path across the hearths and downward, it meets the rising gas stream that effects the roasting. The calcines are discharged from the bottom hearth.

Flash roasters are used only for oxidizing roasts and are, in effect, multiple-hearth roasters with the central hearths removed. This design came with the realization that much of the oxidizing takes place as the particles are actually dropping from hearth to hearth.

Chlorinators are used for roasting oxides to chlorides. They are tall, circular steel shells lined with refractory brick to prevent chlorine attack on the steel. The top of each chlorinator has a sealed hopper for periodic feed charging, and gaseous or liquid chlorine is added at the bottom of the unit. Heat is supplied by electrical resistance through the shell wall and by any exothermic reaction that may occur. The product depends on the chloridizing reaction taking place, with magnesium dichloride, for example, forming as a watery liquid and titanium tetrachloride coming off as a gas.

Calcination of carbonates to oxides is done in a horizontal rotary kiln, which is a mild-steel circular shell lined with refractory material and having a length 10 to 12 times the diameter. Sloping slightly downward from feed to discharge ends, the kiln slowly rotates while fuel-fired burners located inside the kiln provide the required heat.

A sintering machine, or blast roaster, can conduct oxidizing or reducing roasts and then agglomerate the roasted calcines, or it can be used for agglomeration alone. (Agglomeration is the fusing of fine feed material into larger chunks that can be fed into a blast furnace or retort, eliminating the problem of losing the fine feed in the hot air blast). The oxidizing or reducing reaction is exothermic, but in order for agglomeration alone to be conducted, a fuel such as fine coke must be mixed with the charge.

The sintering machine consists of an endless belt of moving metal pallets with grate bottoms on which a fine feed charge is spread and passed under a burner. As the charge ignites, the pallet passes over a suction wind box, so that air being drawn through the feed layer causes combustion (*i.e.,*oxidation) of sulfur or carbon to continue from top to bottom. Because the temperature is high and there is no agitation of the feed, a partial fusion takes place on the surface of the particles, leaving them adhering together in the form of a porous, cellular clinker known as sinter.

Smelting

Smelting is a process that liberates the metallic element from its compound as an impure molten metal and separates it from the waste rock part of the charge, which becomes a molten slag. There are two types of smelting, reduction smelting and matte smelting. In reduction smelting, both the metallic charge fed into the smelter and the slag formed from the process are oxides; in matte smelting, the slag is an oxide while the metallic charge is a combination of metallic sulfides that melt and recombine to give a homogeneous metallic sulfide called matte.

Reduction Smelting

Many types of furnace are used for reduction smelting. The blast furnace is universally used in the reduction of such compounds as iron oxide, zincoxide, and lead oxide, though there are great differences between the furnace designs used in each case. Iron, found naturally in the oxide ores hematite and magnetite, is smelted in a tall, circular, sealed blast furnace. A sintered or pelletized feed consisting of coke (for fuel), limestone (as a flux for slag making), and iron oxide is charged into the top of the furnace through a double bell or rotating chute, and heated air is blown in through nozzles, or tuyeres, close to the furnace bottom. In the ensuing combustion reaction, oxygen in the air combines with carbon in the coke, generating enough heat to melt the furnace charge and forming carbon monoxide, which, in turn, reduces the iron oxide to metallic iron. The furnace is sealed to prevent the escape of carbon monoxide gas, which is recovered and burned as fuel to heat the tuyere air. In the hearth at the furnace bottom, molten slag and iron collect in two layers, the lighter slag on top. Both are periodically drawn off, with the slag being discarded and the iron going on to be refined into steel.

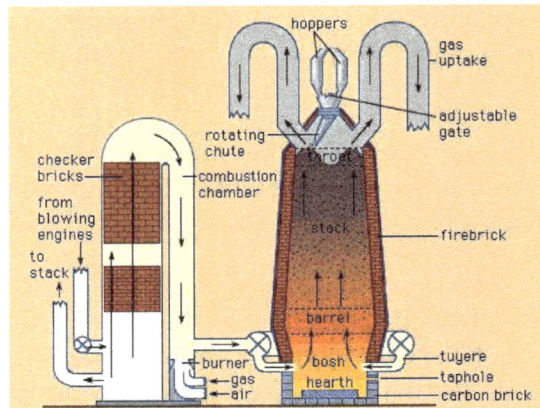

Schematic diagram of modern blast furnace (right) and hot-blast stove (left).

The zinc blast furnace also is a sealed furnace, with a charge of sintered zinc oxide and preheated coke added through a sealed charging bell. The furnace is rectangular, with a shorter shaft than the iron blast furnace. A blast of hot air through the tuyeres provides oxygen to burn the coke for heat and to supply carbon monoxide reducing gas. The reduced zinc passes out of the furnace as vapour, and this is drawn off to a spray bath of molten lead and condensed to liquid zinc metal. Slag and any lead present in the charge are tapped as liquids from the furnace hearth.

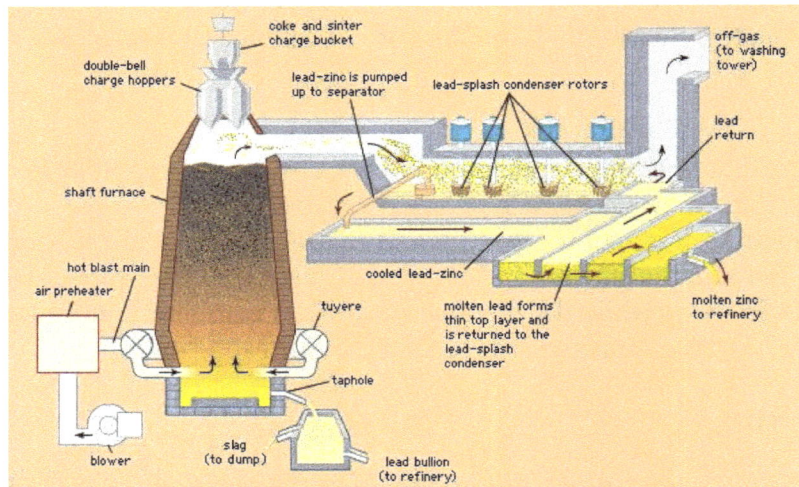

A zinc-lead blast furnace and lead-splash condenser.

The lead blast furnace is similar in size and shape to the zinc blast furnace, but it is not a sealed furnace, and it does not use preheated tuyere air. A charge of lead oxide sinter, coke, and flux is poured into the open top of the furnace, and the strong reducing atmosphere in the furnace shaft reduces the oxide to metal. Liquid lead and slag collect in two layers in the furnace hearth, with the lead in the bottom layer and slag above.

Two newer processes for the direct reduction of unroasted lead sulfide concentrate are the QSL (Queneau-Schuhmann-Lurgi) and the KIVCET (a Russian acronym for "flash-cyclone-oxygen-electric smelting"). In the QSL reactor a submerged injection of shielded oxygen oxidizes lead sulfide to lead metal, while the KIVCET is a type of flash-smelting furnace in which fine, dried lead sulfide concentrate combines with oxygen in a shaft to give lead metal.

Matte Smelting

The primary purpose of matte smelting is to melt and recombine the charge into a homogeneous matte of metallic copper, nickel, cobalt, and iron sulfides and to give an iron and silicon oxide slag. It is done in many types of furnace on both roasted or unroasted sulfide feed material.

The reverberatory furnace is essentially a rectangular refractory-brick box equipped with end-wall burners to provide heat for melting. The furnace is relatively quiet, and it does not blow out much fine feed (which is added through roof ports) with the exhaust gases. The matte is tapped periodically from a centre taphole, while the slag runs off continuously at the furnace flue end. Oxygen lances inserted through the roof, or oxygen added through the burners, can increase smelting capacity considerably.

Electric furnaces are similar to reverberatory furnaces except for the method of heating—in this case a row of electrodes projecting through the roof into the slag layer on the furnace hearth and heating by resistance.

Flash smelting is a relatively recent development that has found worldwide acceptance. It is an autogenous process, using the oxidation of sulfides in an unroasted charge to supply the heat required to reach reaction temperatures and melt the feed material. The most widely used furnace has a vertical reaction shaft at one end of a long, low settling hearth and a vertical gas-uptake shaft at the other end. Fine, unroasted feed is blown into the reaction shaft along with preheated air; these react instantaneously, and liquid droplets fall onto the settling hearth, separating into layers of slag and matte. The off-gas, high in sulfur dioxide, is ideal for sulfur-recovery processes.

The second stage of matte smelting is converting the sulfides to metal. For many years the standard vessel for this operation has been the Peirce-Smith converter. This is a rotatable, refractory-lined, horizontal steel drumwith an opening at the centre of the top for charging and discharging and a row of tuyeres across the back through which air, oxygen-enriched air, or oxygen can be blown into the liquid bath. Molten matte from the smelting furnace is poured into the converter, after which gas is blown through the tuyeres to oxidize first iron and then sulfur. The sulfur goes off as sulfur dioxide gas and the iron as iron oxide slag, leaving semipure metal. Considerable heat is generated by this exothermic reaction, keeping the bath liquid and maintaining the required reaction temperature.

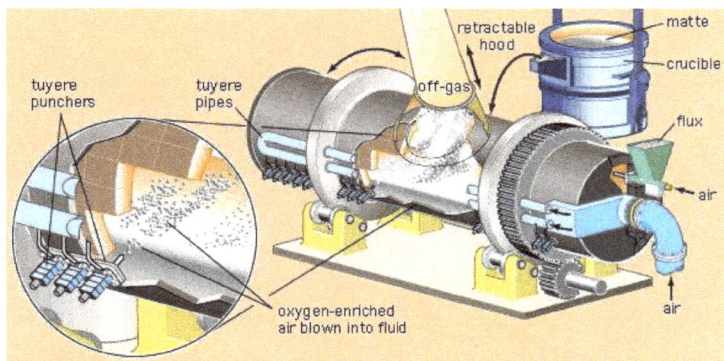

A side-blown copper-nickel matte converter.

More recent processes take advantage of exothermic heat evolution to accomplish both the smelting of unroasted sulfides and the conversion of matte in one combined operation. These are the

Noranda, TBRC (top-blown rotary converter), and Mitsubishi processes. The Noranda reactor is a horizontal cylindrical furnace with a depression in the centre where the metal collects and a raised hearth at one end where the slag is run off. Pelletized unroasted sulfide concentrate is poured into the molten bath at one end, where tuyeres inject an air-oxygen mixture. This causes an intense mixing action that aids the melting, smelting, and oxidation steps, which follow one another in sequence, by taking advantage of the exothermic heat. The TBRC also is cylindrical in shape but is inclined at 17° to the horizontal, has an open mouth at the high end for charging and pouring, and revolves at 5 to 40 rotations per minute. A lance inserted through the mouth can give any combination of oxygen, air, or natural gas to impinge on the molten bath and create the conditions required for smelting and oxidizing. The combination of surface blowing and bath rotation improves the performance of the converter. The Mitsubishi process is a continuous smelting-converting operation that uses three stationary furnaces in series. The first furnace is for smelting, with oxygen lances and a fuel-fired burner inserted through the roof. Slag and matte flow from here to a slag-cleaning furnace (heated by electric arc), and high-grade matte flows from this to the converting furnace, where oxygen-enriched air is blown into the bath through roof lances. Exothermic heat produced here is sufficient to keep the bath up to reaction temperature.

Electrolytic Smelting

Smelting is also carried out by the electrolytic dissociation, at high temperatures, of a liquid metallic chloride compound (as is done with magnesium) or of a metallic oxide powder dissolved in molten electrolyte(as is done with aluminum). In each case, electric current is passed through the bath to dissociate the metallic compound; the metal released collects at the cathode, while a gas is given off at the anode.

The magnesium smelting cell consists of a steel pot that serves as the cathode; two rows of graphite electrodes are inserted through a refractory cover as anodes. The electrolyte is a mixture of chlorides, with magnesium chloride making up 20 percent, and the cell is maintained at 700 °C (1,300 °F). The passage of current breaks down the magnesium chloride into chlorine gas and magnesium metal, which go to the anode and cathode, respectively.

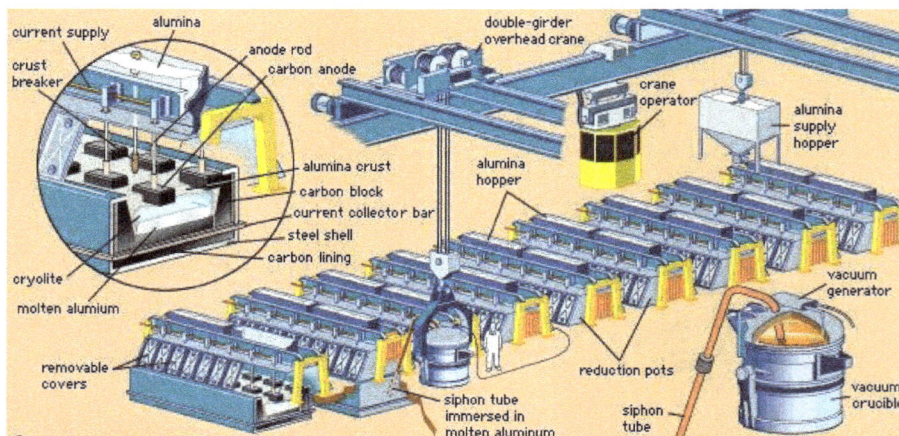

Part of a modern potline based on the electrolytic Hall-Héroult smelting process.

In the Hall-Héroult smelting process, a nearly pure aluminum oxidecompound called alumina is dissolved at 950 °C (1,750 °F) in a molten electrolyte composed of aluminum, sodium, and fluorine; this is electrolyzed to give aluminum metal at the cathode and oxygen gas at the anode. The smelting cell is a carbon-lined steel box, which acts as the cathode, and a row of graphite electrodes inserted into the bath serves as anodes.

Refining

Refining is the final procedure for removing (and often recovering as by-products) the last small amounts of impurities left after the major extraction steps have been completed. It leaves the major metallic element in a practically pure state for commercial application. The procedure is accomplished in three ways: refining by fire, by electrolytic, or by chemical methods.

Fire Refining

Iron, copper, and lead are fire-refined by selective oxidation. In this process, oxygen or air is added to the impure liquid metal; the impurities oxidize before the metal and are removed as an oxide slag or a volatile oxide gas.

The basic oxygen furnace (BOF) is a vessel used to convert pig iron, of about 94 percent iron and 6 percent combined impurities such as carbon, manganese, and silicon, into steel with as little as 1 percent combined impurities. The BOF is a large pear-shaped unit that can be tilted to charge and pour. Molten blast-furnace iron and steel scrap are charged into the furnace; then it is turned to an upright position and a lance inserted to blow high-tonnage oxygen gas into the bath. Oxidation reactions occur rapidly, with silicon and manganese oxidizing first and combining to form an oxide slag, then carbon oxidizing to carbon monoxide gas and burning to carbon dioxide as it leaves the furnace mouth. These reactions are strongly exothermic and keep the vessel up to its reaction temperature without any external heat or fuel being added.

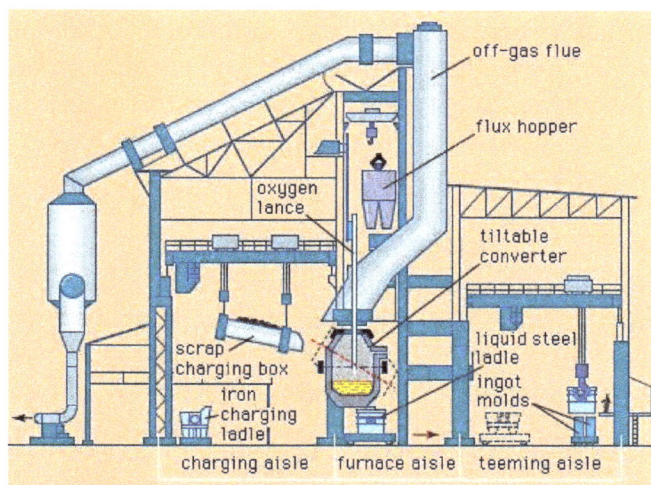

A basic oxygen furnace shop.

Converter-produced blister copper and blast-furnace lead also are treated by fire refining, with both processes depending on the weaker affinity for oxygen of the metals than the impurities they contain. Molten copper in a small reverberatory-type furnace has compressed air blown into it through steel pipes below the surface. This oxidizes zinc, tin, iron, lead, arsenic, antimony, and sulfur; the sulfur goes off as sulfur dioxide gas, while the other impurities form an oxide slag that is skimmed off. Lead is refined in much the same way, with compressed air blown into a molten lead bath and the major impurities of tin, antimony, and arsenic oxidizing in that order, rising to the surface as skims and being scraped off.

Other fire-refining operations use fractional distillation. By this method, zinc metal of 98 percent purity can be upgraded to 99.995 percent purity. The main impurities in blast-furnace zinc are lead and cadmium, with lead boiling at 1,744 °C (3,171 °F), zinc at 907 °C (1,665 °F), and cadmium at 765 °C (1,409 °F). In the first stage zinc and cadmium are boiled off, leaving liquid lead, and in the second stage cadmium is boiled off to leave special high-purity zinc metal.

Electrolytic Refining

This method gives the highest-purity metal product as well as the best recovery of valuable impurities. It is used for copper, nickel, lead, gold, and silver. The metal to be refined is cast into a slab, which becomes the anode of an electrolytic cell; another sheet of metal is the cathode. Both electrodes are immersed in an aqueous electrolyte capable of conducting an electric current. As a direct current is impressed on the cell, metal ions dissolve from the anode and deposit at the cathode. The insoluble sludge left in the cell is treated to recover any valuable by-product metals.

Chemical Refining

An example of chemical refining is the nickel carbonyl process, in which impure nickel metal is selectively reacted with carbon monoxide gas to form nickel carbonyl gas. This gas is then decomposed to give high-purity nickel metal.

Hydrometallurgy

Hydrometallurgy is concerned with the selective leaching of metallic compounds to form a solution from which the metals can be precipitated and recovered. Leaching processes are used when it is the simplest method or when the ore is of too low a grade for more expensive extractive procedures.

Conversion

Because not all ores and concentrates are found naturally in a form that is satisfactory for leaching, they must often be subjected to preliminary operations. For example, sulfide ores, which are relatively insoluble in sulfuric acid, can be converted to quite soluble forms by oxidizing or sulfatizing roasts. On the other hand, oxide ores and concentrates can be given a controlled reducing roast in order to produce a calcine containing a reduced metal that will dissolve easily in the leaching solution.

A second popular treatment for converting sulfides is pressure oxidation, in which the sulfides are oxidized to a porous structure that provides good access for the leaching solution. This treatment was developed for the recovery of gold from sulfide ores, which are not suitable for cyanide leaching without first being oxidized. A finely ground concentrate slurry is preheated to 175 °C (350 °F) and pumped into a four- or five-compartment autoclave, each compartment containing an agitator. Gaseous oxygen is added to each compartment, and retention time in the autoclave is two hours in order to achieve the desired oxidation.

Leaching

Oxides are leached with a sulfuric acid or sodium carbonate solvent, while sulfates can be leached with water or sulfuric acid. Ammonium hydroxide is used for native ores, carbonates, and sulfides, and sodium hydroxide is used for oxides. Cyanide solutions are a solvent for the precious metals, while a sodium chloride solution dissolves some chlorides. In all cases the leach solvent should be cheap and available, strong, and preferably selective for the values present.

Leaching is carried out by two main methods: simple leaching at ambient temperature and atmospheric pressure; and pressure leaching, in which pressure and temperature are increased in order to accelerate the operation. The method chosen depends on the grade of the feed material, with richer feed accommodating a costlier, more extensive treatment.

Leaching in-place, or in situ leaching, is practiced on ores that are too far underground and of too low a grade for surface treatment. A leach solution is circulated down through a fractured ore body to dissolve the values and is then pumped to the surface, where the values are precipitated.

Heap leaching is done on ores of semilow grade—that is, high enough to be brought to the surface for treatment. This method is increasing in popularity as larger tonnages of semilow-grade ore are mined. The ore is piled in heaps on pads and sprayed with leach solution, which trickles down through the heaps while dissolving the values. The pregnant solution is drained away and taken to precipitation tanks.

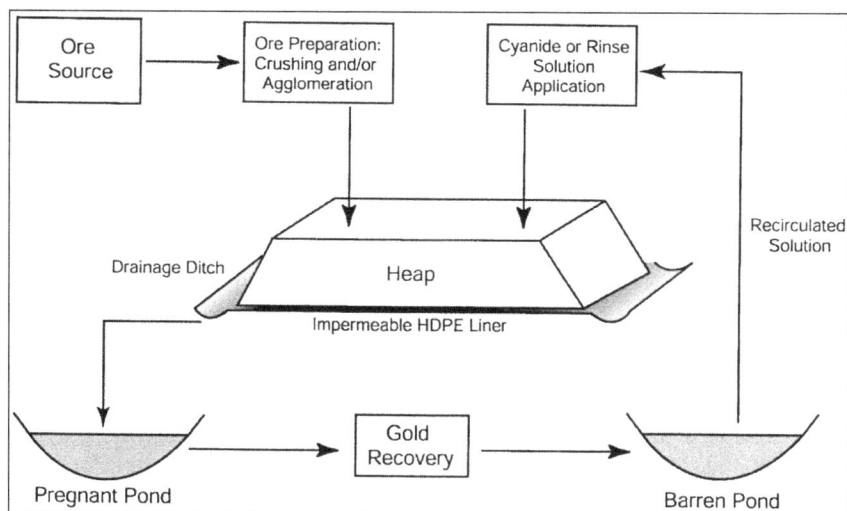

Higher-grade ores are treated by tank leaching, which is carried out in two ways. One method is of very large scale, with several thousand tons of ore treated at a time in large concrete tanks with a circulating solution. In the second method, small amounts of finely ground high-grade ore are agitated in tanks by air or by mechanical impellers. Both solutions pass to precipitation after leaching is completed.

Pressure leaching shortens the treatment time by improving the solubility of solids that dissolve only very slowly at atmospheric pressure. For this process autoclaves are used, in both vertical and horizontal styles. After leaching, the pregnant solution is separated from the insoluble residue and sent to precipitation.

Recovery

Pregnant solution from leaching operations is treated in a variety of ways to precipitate the dissolved metal values and recover them in solid form. These include electrolytic deposition, transfer of metal ions, chemical precipitation, solvent extraction in combination with electrolytic and chemical methods, and carbon adsorption combined with electrolytic treatment.

Electrolytic deposition, also called electrowinning, gives a pure product and is a preferred method. However, it is expensive, owing to the cost of electricity, and must have a solution of high metal content. Insoluble anodes, and cathodes made of either a strippable inert material or a thin sheet of the deposited metal, are inserted into a tank containing leach solution. As current is passed, the solution dissociates, and metal ions deposit at the cathode. This common method is used for copper, zinc, nickel, and cobalt.

Solvent extraction combined with electrolytic deposition takes dilute, low-value metal solutions and concentrates them into small volumes and high metal contents, rendering them satisfactory for electrolytic treatment. Low-grade copper ores are processed in this manner. First, a large volume of a low-value copper leach solution (2.5 grams per litre, or 0.33 ounces per gallon) is contacted with a small volume of water-immiscible organic solvent in kerosene. The metal values pass from the leach solution into the extraction solution, the two phases are separated, and the extraction solution goes on to the stripping circuit. Here another fluid is added that has a still greater affinity for the metal values, picking them out of the extraction solution. The two solutions are separated, with the small volume of stripping solution having a metal content high enough (50 grams per litre, or 6.6 ounces per gallon) to be suitable for electrolytic precipitation.

An adsorption circuit is used to strip pregnant solutions of gold cyanide with activated carbon. The carbon is in turn stripped of the metal by a solution, which then goes to an electrolytic cell where the gold content is deposited at the cathode.

Chemical precipitation can be accomplished in a number of ways. In one method, a displacement reaction takes place in which a more active metal replaces a less active metal in solution. For example, in copper cementation iron replaces copper ions in solution, solid particles of copper precipitating while iron goes into solution. This is an inexpensive method commonly applied to weak, dilute leach solutions. Another displacement reaction uses gas, with hydrogen sulfide, for example, added to a solution containing nickel sulfate and precipitating nickel sulfide. Finally, changing the acidity of a solution is a common method of precipitation. Yellow cake, a common name for sodium diuranate, is precipitated from a concentrated uranium leach solution by adding sodium hydroxide to raise the pH to 7.

Physical Metallurgy

Physical metallurgy is the science of making useful products out of metals. Metal parts can be made in a variety of ways, depending on the shape, properties, and cost desired in the finished product. The desired properties may be electrical, mechanical, magnetic, or chemical in nature; all of them can be enhanced by alloying and heat treatment. The cost of a finished part is often determined more by its ease of manufacture than by the cost of the material. This has led to a wide variety of ways to form metals and to an active competition among different forming methods, as well as among different materials. Large parts may be made by casting. Thin products such as automobile fenders are made by forming metal sheets, while small parts are often made by powder metallurgy (pressing powder into a dieand sintering it). Usually a metal part has the same properties throughout. However, if only the surface needs to be hard or corrosion-resistant, the desired performance can be obtained through a treatment that changes only the composition and strength of the surface.

Structures and Properties of Metals

Metallic Crystal Structures

Metals are used in engineering structures (e.g., automobiles, bridges, pressure vessels) because, in contrast to glass or ceramic, they can undergo appreciable plastic deformation before breaking. This plasticity stems from the simplicity of the arrangement of atoms in the crystals making up a piece of metal and the nondirectional nature of the bond between the atoms. Atoms can be arranged in many different ways in crystalline solids, but in metals the packing is in one of three simple forms. In the most ductile metals, atoms are arranged in a close-packed manner. If atoms were visualized as identical spheres and if these spheres were packed into planes in the closest possible manner, there would be two ways to stack close-packed planes one above another. (One would lead to a crystal with hexagonal symmetry (called hexagonal close-packed, or hcp); the other would lead to a crystal with cubic symmetry that could also be visualized as an assembly of cubes with atoms at the corners and at the centre of each face (known as face-centred cubic, or fcc). Examples of metals with the hcp type of structure are magnesium, cadmium, zinc, and alpha titanium. Metals with the fcc structure include aluminum, copper, nickel, gamma iron, gold, and silver.

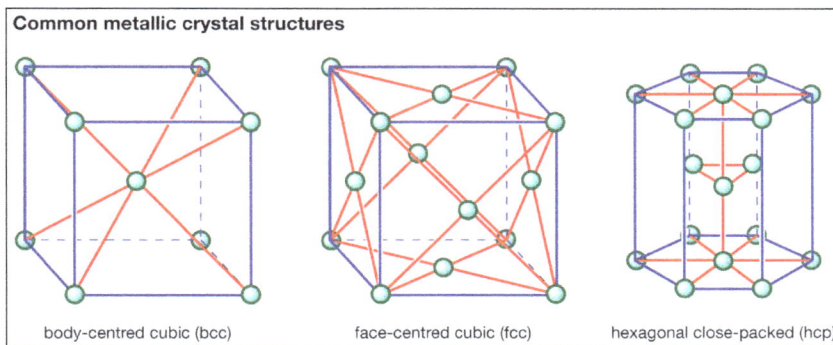

The commonest metallic crystal structures.

The third common crystal structure in metals can be visualized as an assembly of cubes with atoms at the corners and an atom in the centre of each cube; this is known as body-centred cubic, or bcc. Examples of metals with the bcc structure are alpha iron, tungsten, chromium, and beta titanium.

Some metals, such as titanium and iron, exhibit different crystal structures at different temperatures. The lowest-temperature structure is labeled alpha (α), and higher-temperature structures beta (β), gamma (γ), and delta (δ). This allotropy, or transformation from one structure to another with changing temperature, leads to the marked changes in properties that can come from heat treatment.

When a metal undergoes a phase change from liquid to solid or from one crystal structure to another, the transformation begins with the nucleation and growth of many small crystals of the new phase. All these crystals, or grains, have the same structure but different orientations, so that, when they finally grow together, boundaries form between the grains. These boundaries play an important role in determining the properties of a piece of metal. At room temperature they strengthen the metal without reducing its ductility, but at high temperatures they often weaken the structure and lead to early failure. They can be the site of localized corrosion, which also leads to failure.

Mechanical Properties

When a metal rod is lightly loaded, the strain (measured by the change in length divided by the original length) is proportional to the stress This means that, with each increase in load, there is a proportional increase in the rod's length, and, when the load is removed, the rod shrinks to its original size. The strain here is said to be elastic, and the ratio of stress to strain is called the elastic modulus. If the load is increased further, however, a point called the yield stress will be reached and exceeded. Strain will now increase faster than stress, and, when the sample is unloaded, a residual plastic strain (or elongation) will remain. The elastic strain at the yield stress is typically 0.1 to 1 percent, whereas, with the sample pulled to rupture, the plastic strain is typically 20 to 40 percent for an alloy (it may exceed 100 percent in some cases).

The most important mechanical properties of a metal are its yield stress, its ductility (measured by the elongation to fracture), and its toughness(measured by the energy absorbed in tearing the metal). The yield stress of a metal is determined by the resistance to slipping of one plane of atoms over another. Various barriers to slip can be produced by heat treatment and alloying; examples of such barriers are grain boundaries, fine precipitates, distortion introduced by cold working the metal, and alloying elements dissolved in the metal.

When a metal is made very strong through one or more of these methods, it may suddenly fracture under a load instead of yielding. This is particularly true when the metal contains notches or cracks that locally raise the stress and localize the yielding. The property of interest then becomes the fracture toughness, measured by the energy required to extend an existing crack in a piece of metal. In almost all cases, the fracture toughness of an alloy can be improved only by reducing its yield strength. The only exception to this is a smaller grain size, which increases both toughness and strength.

Electrical Properties

The electrical conductivity of a metal (or its reciprocal, electrical resistivity) is determined by the ease of movement of electrons past the atoms under the influence of an electric field. This movement is particularly easy in copper, silver, gold, and aluminum—all of which are well-known conductors of electricity. The conductivity of a given metal is decreased by phenomena that deflect, or scatter, the moving electrons. These can be anything that destroys the local perfection of the atomic arrangement—for example, impurity atoms, grain boundaries, or the random oscillation of atoms induced by thermal energy. This last example explains why the conductivity of a metal increases substantially with falling temperature: in a pure metal at room temperature, most resistance to the motion of free electrons comes from the thermal vibration of the atoms; if the temperature is reduced to almost absolute zero, where thermal motion essentially stops, conductivity can increase several thousandfold.

Magnetic Properties

When an electric current is passed through a coil of metal wire, a magnetic field is developed around the coil. When a piece of copper is placed inside the coil, this field increases by less than 1 percent, but, when a piece of iron, cobalt, or nickel is placed inside the coil, the external field can increase 10,000 times. This strong magnetic property is known as ferromagnetism, and the three metals listed above are the most prominent ferromagnetic metals. When the piece of ferromagnetic metal is removed from the coil, it retains some of this magnetism (that is, it is magnetized). If the metal is hard, as in a hardened piece of steel, the loss, or reversal, of magnetizationwill be slow, and the sample will be useful as a permanent magnet. If the metal is soft, it will quickly lose its magnetism; this will make it useful in electrical transformers, where rapid reversal of magnetization is essential.

In many types of solids, the atoms possess a permanent magnetic moment(they act like small bar magnets). In most solids, the direction of these moments is arranged at random. What is exceptional about ferromagnetic solids is that the interatomic forces cause the moments of neighbouring atoms spontaneously to align in the same direction. If the moments of all of the atoms in a single sample lined up in the same direction, the sample would be an exceptionally strong magnet with exceptionally high energy. That energy would be reduced if the sample broke up into domains, with all atomic moments in each domain being aligned but the direction of magnetization in adjacent domains being in opposite directions and thus tending to cancel one another. This is what happens when a ferromagnetic metal is magnetized: all domains do not take on the same orientation, but domains of one orientation grow at the expense of others. The alignment of atomic magnetic moments within a domain is weakened by thermally induced oscillations, and ferromagnetism is finally lost above the Curie point, which is 770 °C (1,420 °F) for iron and 358 °C (676 °F) for nickel.

Chemical Properties

Almost any metal will oxidize in air, the only exception being gold. At room temperature a clean metal surface will oxidize very little, since a thin oxidefilm forms and protects the metal from further oxidation. At elevated temperatures, though, oxidation is faster, and the film is less protective. Many chemicals accelerate this corrosion process (that is, the conversion of a metal to an oxide in air or to a hydroxide in the presence of water).

A special property of metal surfaces is their ability to catalyze chemical reactions. For example, in the exhaust system of most automobiles, combustion gases pass over a dispersion of very fine platinum particles. The surfaces of these particles greatly accelerate the oxidation of carbon monoxide and hydrocarbons to carbon dioxide and water, thus reducing the toxicity of the exhaust gases.

Alloying

Almost all metals are used as alloys—that is, mixtures of several elements—because these have properties superior to pure metals. Alloying is done for many reasons, typically to increase strength, increase corrosion resistance, or reduce costs.

Processes

In most cases, alloys are mixed from commercially pure elements. Mixing is relatively easy in the liquid state but slow and difficult in the solid state, so that most alloys are made by melting the base metal—for instance, iron, aluminum, or copper—and then adding the alloying agents. Care must be taken to avoid contamination, and in fact purification is often carried out at the same time, since this is also done more easily in the liquid state. Examples can be found in steelmaking, including the desulfurizing of liquid blast-furnace iron in a ladle, the decarburization of the iron during its conversion to steel, the removal of oxygen from the liquid steel in a vacuum degasser, and finally the addition of tiny amounts of alloying agents to bring the steel to the desired composition.

The largest tonnages of alloys are melted in air, with the slag being used to protect the metal from oxidation. However, a large and increasing amount is melted and poured entirely in a vacuum chamber. This allows close control of the composition and minimizes oxidation. Most of the alloying elements needed are placed in the initial charge, and melting is done with electricity, either by induction heating or by arc melting. Induction melting is conducted in a crucible, while in arc melting the melted droplets drip from the arc onto a water-cooled pedestal and are immediately solidified.

Sometimes an inhomogeneous, composite structure is desired, as in cemented tungsten carbide cutting tools. In such cases, the alloy is not melted but is made by powder metallurgical techniques.

Metallurgy

Increasing Strength

The most common reason for alloying is to increase the strength of a metal. This requires that barriers to slip be distributed uniformly throughout the crystalline grains. On the finest scale, this is done by dissolving alloying agents in the metal matrix (a procedure known as solid solution hardening). The atoms of the alloying metals may substitute for matrix atoms on regular sites (in which case they are known as substitutionalelements), or, if they are appreciably smaller than the matrix atoms, they may take up places between regular sites (where they are called interstitial elements).

The next coarser type of barrier to slip is a fine, solute-rich precipitate with dimensions of only tens or hundreds of atomic diameters. These particles are formed by heat treatment. The metal is heated to a temperature at which the solute-rich phase dissolves (e.g., 5 percent copper in aluminum at 540 °C [1,000 °F]), and then it is rapidly cooled to avoid precipitation. The next step is to form a fine precipitate throughout the sample by aging at an elevated temperature that is well below the temperature used for the initial dissolution.

In metals that undergo transformations from one crystal structure to another on heating (e.g., iron or titanium), the difference in solute solubility between the high- and low-temperature phases is often utilized. For example, in the low-alloy steels used for tools and gears, carbon forms the hardening precipitate. Carbon is much more soluble in the high-temperature fcc phase (gamma iron, also called austenite) than in the low-temperature bcc phase (alpha iron, or ferrite). The other alloying elements added (e.g., chromium, nickel, and molybdenum) retard the transformation of austenite on cooling, so that the fcc-to-bcc transformation occurs at a low temperature by a sudden, shear transformation; this allows no time for carbon precipitation and makes the steel harder. A final reheating tends to coarsen the precipitate and thereby increase ductility; this is commonly called tempering.

An array of barriers on the same scale as precipitation hardening can be created by plastically deforming the metal at room temperature. This is often done in a cold-working operation such as rolling, forging, or drawing. The deformation occurs through the generation and motion of line defects, called dislocations, on slip planes spaced only a few hundred atom diameters apart. When slip occurs on different planes, the intersecting dislocations form tangles that inhibit further slip on those planes. Such strain hardening can double or triple the yield stress of a metal.

Increasing Corrosion Resistance

Alloys can have much better high-temperature oxidation resistance than pure metals. The alloying elements most commonly used for this purpose are chromium and aluminum, both of which form an adherent film of stable oxide on the surface that protects the metal from further oxidation. Eleven percent or more chromium is added to iron to create a stainless steel, while 10 to 15 percent chromium and 3 to 5 percent aluminum are commonly added to the nickel- or cobalt-based superalloys used in the highest-temperature components of jet engines.

Inhibiting the corrosion of alloys in water is more varied and complex than inhibiting high-temperature oxidation. Nevertheless, one of the most common techniques is to add alloying elements that inhibit the corrosion.

Reducing Costs

Gold and silver used in jewelry and coins are alloyed with other metals to increase strength and reduce cost. Sterling silver contains 7.5 percent base metal, commonly copper. The fraction of gold in gold jewelry is designated in karats, with 24-karat being pure gold and 18-karat being 75 percent gold by weight. In coins, alloys with the look and density of silver are commonly substituted for silver; for instance, all U.S. coins that appear to be made of silver actually have a surface layer of 75 percent copper and 25 percent nickel.

Lowering Melting Points

Alloying can also be done to lower the melting point of a metal. For example, adding lead to tin lowers the melting point of the tin-rich alloy, and adding tin to lead lowers the melting point of the lead-rich alloy. A 62-percent-tin 38-percent-lead alloy, which is called the eutectic composition, has the lowest melting point of all, much lower than that of either metal. Eutectic lead-tin alloys are used for soldering.

Casting

Casting consists of pouring molten metal into a mold, where it solidifies into the shape of the mold. The process was well established in the Bronze Age, when it was used to form most of the bronzepieces now found in museums. It is particularly valuable for the economical production of complex shapes, ranging from mass-produced parts for automobiles to one-of-a-kind production of statues, jewelry, or massive machinery.

Processes

Casting processes differ in how the mold is made and in how the metal is forced into the mold. For metals with a high melting temperature, stable refractory material must be used to avoid reaction between the metal and the mold. Most steel and iron castings, for example, are poured into silica sand, though some parts are cast into coated metal molds. For metals of lower melting point, such as aluminum or zinc, molds can be made of another metal or of sand, depending on how many parts are to be produced and other considerations. Gravity is most frequently employed to fill the mold, but some processes use centrifugal force or pressure injection.

Sand-casting

Sand-casting is widely used for making cast-iron and steel parts of medium to large size in which surface smoothness and dimensional precision are not of primary importance.

The first step in any casting operation is to form a mold that has the shape of the part to be made. In many processes, a pattern of the part is made of some material such as wood, metal, wax, or polystyrene, and refractory molding material is formed around this. For example, in green-sand-casting, sand combined with a binder such as water and clay is packed around a pattern to form the mold. The pattern is removed, and on top of the cavity is placed a similar sand mold containing a passage (called a gate) through which the metal flows into the mold. The mold is designed so that solidification of the casting begins far from the gate and advances toward it, so that molten metal in the gate can flow in to compensate for the shrinkage that accompanies solidification. Sometimes additional spaces, called risers, are added to the casting to provide reservoirs to feed this shrinkage. After solidification is complete, the sand is removed from the casting, and the gate is cut off. If cavities are intended to be left in the casting—for example, to form a hollow part—sand shapes called cores are made and suspended in the casting cavity before the metal is poured.

Patterns are also formed for sand-casting out of polymers that are evaporated by the molten metal. Such patterns may be injection molded and can possess a very complex shape. The process is called full-mold or evaporative pattern casting.

A variant of sand-casting is the shell-molding process, in which a mixture of sand and a thermosetting resin binder is placed on a heated metal pattern. The resin sets, binding the sand particles together and forming half of a strong mold. Two halves and any desired cores are then assembled to form the mold, and this mold is backed up with moist sand for casting. Greater dimensional accuracy and a smoother surface are obtained in this process than in greensand-casting.

Metal Molds

Other molds are made from metal. Here a die of the desired shape is machined from cast iron or steel. If the metal flows into the mold by gravity, the process is called permanent mold casting. If the molten metal is forced in under pressure, the process is called die casting. Die-casting dies are water-cooled; consequently, they can produce parts with thinner walls at a higher rate than permanent mold machines. The rapid cooling creates a stronger part than sand-casting, but ductility may be poorer owing to entrapped gas and porosity.

Since the initial cost of a die is substantial, metal molds are cost-effective only when many identical parts are to be made. Indeed, a die may be made to produce several parts at once.

Investment Casting

In investment casting a mold is made by drying a refractory slurry on a pattern made of wax or plastic. A series of layers is applied and dried to make a ceramic shell, and the pattern is then melted or burned out to provide the mold. This process allows the mass production of parts with more complex shapes and finer surface detail than can be attained by other processes. It can be used with almost any metal and is customarily employed for casting relatively small parts. The wax pattern can be made by injection molding.

Centrifugal Casting

Centrifugal casting forces the metal into a mold by spinning it. It is used for the casting of small precious-metal objects, so that essentially all of the metal goes into the casting instead of the gates and risers. It is also used to produce long, hollow objects without resorting to cores—for example, to cast pipe. Here the long, cylindrical mold is horizontal and is spun about the axis of the cylinder as metal is poured into the mold.

Continuous Casting

Actually not a means of casting parts, continuous casting is practiced in the primary production of metals to form strands for further processing. The metal is poured into a short, reciprocating, water-cooled mold and solidifies even as it is withdrawn from the other side of the mold. The process is widely used in the steel industry because it eliminates the cost of reheating ingots and rolling them to the proportions of the billets, blooms, and slabs made by continuous casting.

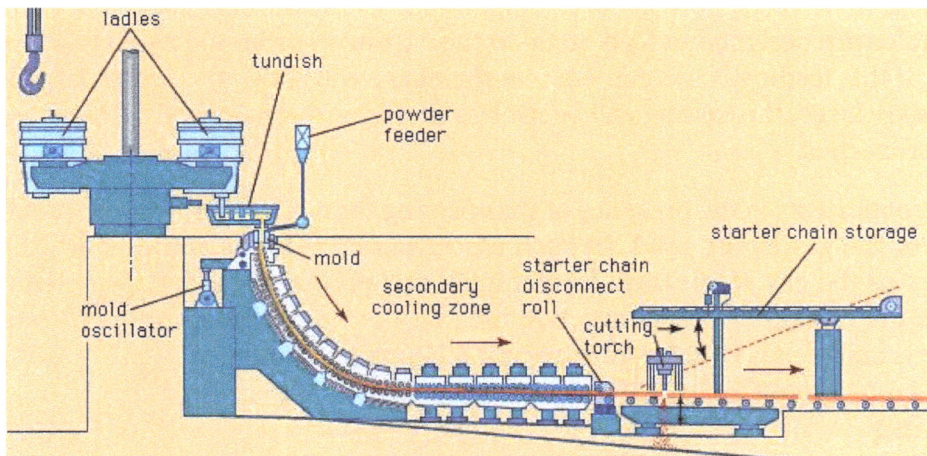

A curved-mold continuous slab caster.

Metallurgy

The mechanical properties of castings can be degraded by inhomogeneities in the solidifying metal. These include segregation, porosity, and large grain size.

Grain Size

A fine-grained casting can be produced by rapidly cooling the liquid metal to well below its equilibrium freezing temperature—*i.e.*, by pouring into a mold that cools the metal rapidly. For this reason, die castings have a finer grain size than the same alloy cast in a sand mold.

In cast iron, remarkable changes in microstructure result from various alloying additions and casting temperatures. For example, normal cast iron solidified in a sand mold forms what is known as gray iron, an iron matrix containing about 20 percent by volume graphite flakes. This type of iron has limited ductility. However, when a small amount of magnesium is added to the melt before casting, the result is a "spheroidal graphite" iron, in which graphite appears as spherical nodules and ductility is greatly increased. If the molten iron is chill cast (*i.e.*, rapidly cooled), it will form a "white" iron containing about 60 percent cementite, or iron carbide. This material is hard and wear-resistant, but it has no ductility at all. These cast irons are usually given a heat treatment to improve their mechanical properties.

Segregation

Different parts of a casting may have different compositions, stemming from the fact that the solid freezing out of a liquid has a different composition from the liquid with which it is in contact. (For example, when salt water is cooled until ice forms, the ice is essentially pure water while the salt concentration of the water rises). Minor segregation is unimportant, but large differences can lead to local spots that are exceptionally weak or strong, and both of these can lead to early failure in a part under stress.

Porosity

A major problem in castings, porosity is principally caused by the shrinkage that accompanies solidification. Molds are designed to feed metal to the casting in order to keep it full as solidification proceeds, but, if this feeding is incomplete, the shrinkage will show up as internal pores or cracks. If these cracks are large, the casting will be useless. If they are small, they will have relatively little effect on the properties.

Another cause of porosity is the presence of gas-forming impurities in the liquid metal that exceed the solubility of the gas in the solid. In such cases, solidification is accompanied by the formation of bubbles as the gas is rejected. To eliminate this problem, gas-forming elements must be removed from the liquid before casting. Bubbling an inert gas such as argon through the liquid before casting is one means of doing this; vacuum degassing is another.

Metalworking

Processes

Metals are important largely because they can be easily deformed into useful shapes. Literally hundreds of metalworking processes have been developed for specific applications, but these can be divided into five broad groups: rolling, extrusion, drawing, forging, and sheet-metal forming. The first four processes subject a metal to large amounts of strain. However, if deformation occurs at a sufficiently high temperature, the metal will recrystallize—that is, its deformed grains will be consumed by the growth of a set of new, strain-free grains. For this reason, a metal is usually rolled, extruded, drawn, of forged above its recrystallization temperature. This is called hot working, and under these conditions there is virtually no limit to the compressive plastic strain to which the metal can be subjected.

Other processes are performed below the recrystallization temperature. These are called cold working. Cold working hardens metal and makes the part stronger. However, there is a definite limit to the strain that can be put into a cold part before it cracks.

Rolling

Rolling is the most common metalworking process. More than 90 percent of the aluminum, steel, and copper produced is rolled at least once in the course of production—usually to take the metal from a cast ingot down to a sheet or bar. The most common rolled product is sheet. With high-speed computer control, it is common for several stands of rolls to be combined in series, with thick sheet entering the first stand and thin sheet being coiled from the last stand at linear speeds of more than 100 kilometres (60 miles) per hour. Similar multistand mills are used to form coils of wire rod from bars. Other rolling mills can press large bars from several sides to form I-beams or railroad rails.

Rolling can be done either hot or cold. If the rolling is finished cold, the surface will be smoother and the product stronger.

Extrusion

Extrusion converts a billet of metal into a length of uniform cross section by forcing the billet to flow through the orifice of a die. In forward extrusion the ram and the die are on opposite sides of the workpiece. Products may have either a simple or a complex cross section; examples of complex extrusions can be found in aluminum window frames.

Tubes or other hollow parts can also be extruded. The initial piece is a thick-walled tube, and the extruded part is shaped between a die on the outside of the tube and a mandrel held on the inside.

In impact extrusion (also called back-extrusion), the workpiece is placed in the bottom of a hole (the die), and a loosely fitting ram is pushed against it. The ram forces the metal to flow back around it, with the gap between the ram and the die determining the wall thickness. When toothpaste tubes were made of a lead alloy, they were formed by this process.

Drawing

Drawing consists of pulling metal through a die. One type is wire drawing. The diameter reduction that can be achieved in such a die is limited, but several dies in series can be used to obtain the desired reduction. Deep drawing starts with a disk of metal and ends up with a cup by pushing the metal through a hole (die). Several drawing operations in sequence may be used for one part. Deep drawing is employed in making aluminum beverage cans and brass rifle cartridges from sheet.

Sheet Metal Forming

In stretch forming, the sheet is formed over a block while the workpiece is under tension. The metal is stretched just beyond its yield point (2 to 4 percent strain) in order to retain the new shape.

Bending can be done by pressing between two dies. Often a part can be made equally well by either stretch forming or bending; the choice then is made on the basis of cost. Shearing is a cutting operation similar to that used for cloth. In these methods the thickness of the sheet changes little in processing.

Each of these processes may be used alone, but often all three are used on one part. For example, to make the roof of an automobile from a flat sheet, the edges are gripped and the piece pulled in tension over a lower die. Next a mating die is pressed over the top, finishing the forming operation, and finally the edges are sheared off to give the final dimensions.

Forging

Forging is the shaping of a piece of metal by pushing with open or closed dies. It is usually done hot in order to reduce the required force and increase the metal's plasticity.

Open-die forging is usually done by hammering a part between two flat faces. It is used to make parts that are too big to be formed in a closed dieor in cases where only a few parts are to be made and the cost of a die is therefore unjustified. The earliest forging machines lifted a large hammer that was then dropped on the work, but now air- or steam-driven hammers are used, since these allow greater control over the force and the rate of forming. The part is shaped by moving or turning it between blows. A forged ring can be formed by placing a mandrel through the ring and deforming the metal between the hammer and the mandrel. Rings also can be forged by rolling with one roll inside the ring and the other outside.

Closed-die forging is the shaping of hot metal within the walls of two dies that come together to enclose the workpiece on all sides. The process starts with a rod or bar cut to the length needed to fill the die. Since large, complex shapes and large strains are involved, several dies may be used to go from the initial bar to the final shape. With closed dies, parts can be made to close tolerances so that little finish machining is required.

Two closed-die forging operations given special names are upsetting and coining. Coining takes its name from the final stage of forming metal coins, where the desired imprint is formed on a smooth metal disk that is pressed in a closed die. Coining involves small strains and is done cold to enhancesurface definition and smoothness. Upsetting involves a flow of the metal back upon itself. An example of this process is the pushing of a short length of a rod through a hole, clamping the rod, and then hitting the exposed length with a die to form the head of a nail or bolt.

Metallurgy

An important benefit of hot working is that it provides control over and improvement of mechanical properties. Hot-rolling or hot-forging eliminate much of the porosity, directionality, and segregation that may be present in cast shapes. The resulting "wrought" product therefore has better ductilityand toughness than the unworked casting. During the forging of a bar, the grains of the metal become greatly elongated in the direction of flow. As a result, the toughness of the metal is substantially improved in this direction and somewhat weakened in directions transverse to the flow. Part of the design of a good forging is to assure that the flow lines in the finished part are oriented so as to lie in the direction of maximum stress when the part is placed in service.

The ability of a metal to resist thinning and fracture during cold-working operations plays an important role in alloy selection and process design. In operations that involve stretching, the best alloys are those which grow stronger with strain (strain harden)—for example, the copper-zinc alloy, brass, used for cartridges and the aluminum-magnesium alloys in beverage cans, which exhibit greater strain hardening than do pure copperor aluminum, respectively.

Another useful property that can be controlled by processing and composition is the plastic anisotropy ratio. When a segment of sheet is strained (*i.e.,* elongated) in one direction, the thickness and width of the segment must shrink, since the volume remains constant. In an isotropic sheet the thickness and width show equal strain, but, if the grains of the sheet are oriented properly, the thickness will shrink only about half as much as the width. Since it is thinning that leads to early fracture, this plastic anisotropy imparts better deep-drawing properties to sheet material with the optimum grain orientation.

Fracture of the workpiece during forming can result from flaws in the metal; these often consist of nonmetallic inclusions such as oxides or sulfides that are trapped in the metal during refining. Such inclusions can be avoided by proper manufacturing procedures. Laps are another type of flaw in which part of a metal piece is inadvertently folded over on itself but the two sides of the fold are not completely welded together. If a force tending to open this fold is applied during the forming operation, the metal will fail at the lap.

The ability of different metals to undergo strain varies appreciably. The shape change that can be made in one forming operation is often limited by the tensile ductility of the metal. Metals with the face-centred cubic crystal structure, such as copper and aluminum, are inherently more ductile in such operations than metals with the body-centred cubic structure. To avoid early fracture in the latter type of metals, processes are used that apply primarily compressive stresses rather than tensile stresses.

Powder Metallurgy

Powder metallurgy (P/M) consists of making solid parts out of metalpowders. The powder is mixed with a lubricant, pressed into a die to form the desired shape, and then sintered, or heated to a temperature below the melting point of the alloy where solid-state bonding of the particles takes place. In the absence of any external force, sintering typically leaves the sample containing about 5 percent pores by volume, but, when pressure is applied during sintering (a process called hot pressing), virtually zero porosity remains. In some parts made by mixing two different elements, one component melts at the sintering temperature, and this liquid phase aids sintering of the solid particles.

Applications

The following roughly chronological account indicates the types of products that can be made by P/M.

The earliest commercial use of P/M was in the production of such high-melting-point metals as platinum, tungsten, and tantalum. Pure powders of these metals could be made by the low-temperature reduction of powders, usually oxides, and, since these metals melt at extremely high temperatures, it was easier to form solid parts by pressing and sintering the powders than by melting and casting. For example, P/M played an important role in the development of tungsten filaments for electric light bulbs.

Another early P/M product was porous-metal bearings and filters. In such parts sintering is conducted at a relatively low temperature so that the pores between the particles remain open and connected. Disks sintered in this way can serve as filters for liquids, or the sintered part can be impregnated with oil to make a self-lubricating bearing. In the latter case, the oil is held in the pores by surface tension. When the bearing heats up in use, some oil flows out and lubricates the surface, and, when the part cools, surface tension pulls the oil back into the fine channels.

Cemented carbides form another class of sintered product. Pure tungsten carbide (WC) is an extremely hard compound, but it is too brittle to be used in tools. However, useful tools can be made by mixing WC powder with cobalt powder and sintering at a temperature above the melting point of cobalt. Liquid cobalt then reacts with the surface of the WC, and when the part is cooled the cobalt freezes, holding the WC tightly together to form a composite structure with enough toughness to be used for tools and dies.

The greatest volume of P/M parts is now produced from iron powder, a process that was first developed during World War II. Small, complex parts, such as gears, require much work if machined from steel bars, and a significant volume of material is wasted as chips from the machining. However, if the part is made by P/M processes, little or no machining is necessary, there is less wasted material, and the cost is much lower. Many small parts for automobiles and appliances are produced in this manner. The second greatest volume of P/M parts is made from aluminum powder. These parts are light, corrosion-resistant, and (if alloy is used) can be heat treated to appreciably increase the strength. Small parts for automobiles and appliances are the most common applications.

A recent process uses P/M methods to improve the homogeneity and toughness of high-alloy tool steels. Cast ingots of these alloys contain a coarse network of brittle phases that are very difficult to break up by hot working, but if, instead of being cast into ingots, the liquid is atomized (solidified as small droplets), the rapidly solidified particles will be homogeneous. This powder can then be hot pressed into consolidated bars with better mechanical properties than those produced by ingot casting. Consolidation is often achieved by hot isostatic pressing, wrapping the pressed powder in an envelope of steel or glass, and heating it in a hot inert gas at high pressure. The consolidated metal is then worked into finished parts.

Processes

The most common method of producing metal powders is atomization of a liquid. Here a stream of molten metal is broken up into small droplets with a jet of water, air, or inert gas such as nitrogen or argon. Atomization in water yields irregularly shaped particles that can be pressed to a higher initial, or "green," strength and density than can the spherical particles formed by atomizing with an inert gas.

In other atomization processes, centrifugal force is used. The metal can be poured onto a spinning disk that breaks up the stream, or a spinning rod can be melted by an electric arc so that it throws off particles as it spins.

The liquid metal being atomized may be an alloy or a pure metal that will subsequently be blended with other elements to form an alloy. After atomization, the powder must be separated into size ranges by passing it through a series of sieves. Powders of different sizes (and of different metals) are then blended for pressing into parts.

Powders are often produced by chemically reducing a powdered oxide—for example, iron oxide reduced with carbon or hydrogen. The resulting metal aggregate is then milled and sieved to obtain the desired powder. Powder also can be made by electrodepositing the metal at a high current density, followed by milling to break up the deposit.

The above processes often produce powders that are roughly 50 to 200 micrometres in diameter. Powders less than one-tenth this size can be found in the finest fraction of the powder produced by atomization. Such fine powders can be mixed with a wax, injection molded to form several parts at once, and then sintered. The resulting parts require very little machining to yield a finished product.

Heat Treating

The properties of metals can be substantially changed by various heat-treatment processes. Depending on the alloy and its condition, heat treating can harden or soften a metal.

Hardening Treatments

Hardening heat treatments invariably involve heating to a sufficiently high temperature to dissolve solute-rich precipitates. The metal is then rapidly cooled to avoid reprecipitation; often this is done by quenching in water or oil. The concentration of solute dissolved in the metal is now much greater than the equilibrium concentration. This produces what is known as solid-solution hardening, but the alloy can usually be hardened appreciably more by aging to allow a very fine precipitate to form. Aging is done at an elevated temperature that is still well below the temperature at which the precipitate will dissolve. If the alloy is heated still further, the precipitate will coarsen; that is, the finest particles will dissolve so that the average particle size will increase. This will reduce the hardness somewhat but increase the ductility. Precipitation hardening is used to produce most high-strength alloys. In products made of soft, ductile metals such as aluminum or copper, the age-hardened alloy is put into service with the finest precipitate (that is, the highest strength) possible.

When heated to a high temperature, a few metals, principally iron and titanium alloys, transform to a different crystal structure. Often the high-temperature phase has a higher solubility for the solute, thus aiding the dissolution of the precipitate. If the alloy is slowly cooled, the reverse phase transformation will occur at a high temperature, forming a coarse precipitate and yielding a soft structure. This is the principle behind annealing procedures. However, if the alloy is quenched from the high temperature, the reverse transformation occurs at a much lower temperature, so that a very fine precipitate forms. This is the basis for hardening iron-carbon (steel) alloys. The hardness of the low-temperature-transformation phase (known as martensite) increases with carbon content, and this can result in some very strong alloys. Other alloying elements such as nickel, chromium, and manganese are added to steelprimarily to slow transformation from the high-temperature phase so that thicker pieces, which cool more slowly on quenching, will harden to martensite on cooling. Steel with fresh martensite is still not tough enough for use without first being heated to an elevated temperature. This tempering process reduces residual stresses produced by the phase transformation, reduces hardness by coarsening the carbide precipitate, and increases toughness. Where high strength is the main concern, the tempering temperature is kept low. When toughness is the primary goal and strength secondary, a relatively high tempering temperature is used.

Softening Treatments

In many situations the purpose of heat treating is to soften the alloy and thereby increase its ductility. This may be necessary if a number of cold-forming operations are required to form a part but the metal is so hardened after the first operation that further cold working will cause it to crack. If the metal is recrystallized by annealing at an elevated temperature, it will become soft enough to allow further forming operations. Another case arises when it is necessary to machine high-carbon tools in order to form a die. If the alloy is quenched from a high temperature to form martensite, it will be hard, brittle, and virtually impossible to machine. But if it is slowly cooled, the carbides will be much coarser, and the steel will be machinable.

Processes

Most furnaces designed for heat treatment use natural gas or electricity to raise the temperature. The atmosphere around the work may be air for low-temperature anneals, but at elevated temperatures some atmosphere other than air must be used in order to avoid oxidation. One common atmosphere is obtained by burning natural gas with less than the stoichiometric amount of air. With more reactive metals, annealing can be done in a vacuum furnace.

In some heat treatments only the surface need be heated. With electromagnetic induction or by use of a laser, this can be done so quickly that no special atmosphere is needed to avoid oxidation. Surface heat treating also avoids the distortion that can accompany heating and quenching the entire part. For example, the rear axle of most automobiles is a steel bar roughly 1 metre long and 3 centimetres in diameter (about 3 feet long and 1.25 inches in diameter). The surface can be hardened by passing the bar through an induction coil that quickly heats the surface immediately beneath the coil to red heat, transforming it to austenite. The inside remains cold, however, and, after the coil passes, this cold interior quickly draws heat from the surface, transforming it to martensite. The part is then tempered and put into service. Since most of the cross section of the part is not heated or transformed during this operation, there is no distortion and therefore no need to straighten a hardened part.

Surface Treating

Because it is the surface of a metal that people see and that reacts with the environment around it, special effort is sometimes made to add lustre, colour, or texture to a surface. In addition, special corrosion-resistant layers are placed on the surface for some applications, and in yet others the surface is hardened to add strength and reduce wear.

Angle grinder Worker using an angle grinder to smooth a piece of metal.

Corrosion Resistance

Barrier Protection

When a metal corrodes in water, the atoms lose electrons and become ions that move into the water. This is called an anodic reaction, and for the corrosion process to proceed there must be a corresponding cathodicreaction that adsorbs the electrons. The process can be stopped by isolating the metal from the water with an impermeable barrier. One of the older applications of this idea is the tin can. Unlike steel, tin is not affected by the acids in food, so that a layer of tin placed on steel sheet protects the steel in the can from corrosion.

The exterior surfaces of many large household appliances consist of steel covered with a layer of coloured glass called enamel. Enamel is inert and adheres tightly to the steel, thus protecting it from corrosion as well as providing an attractive appearance. Decorative chromium plating is another example of a protective-barrier coating on steel. Since chromium does not adhere well to steel, the steel is first electroplated with layers of copper and nickel before being plated with a thin layer of chromium.

The protective layers are metallic, but the most common protective barriers are organic. Paints, polymers, and thin lacquer films are used for various applications near room temperature.

The oxide layer that forms on metals when they are exposed to air also constitutes a protective barrier. Stainless steel and aluminum form the most stable and protective of such films. The thickness of the oxide film on aluminum is often increased by making the part function as the anode in an electrolytic cell. This process, called anodizing, enhances the corrosion resistance somewhat and makes colouring the surface easier. The films that form on copper and steel as a result of corrosion (commonly known as tarnish and rust) are somewhat thicker and show a characteristic colour that is often incorporated into the design of the part.

Galvanic Protection

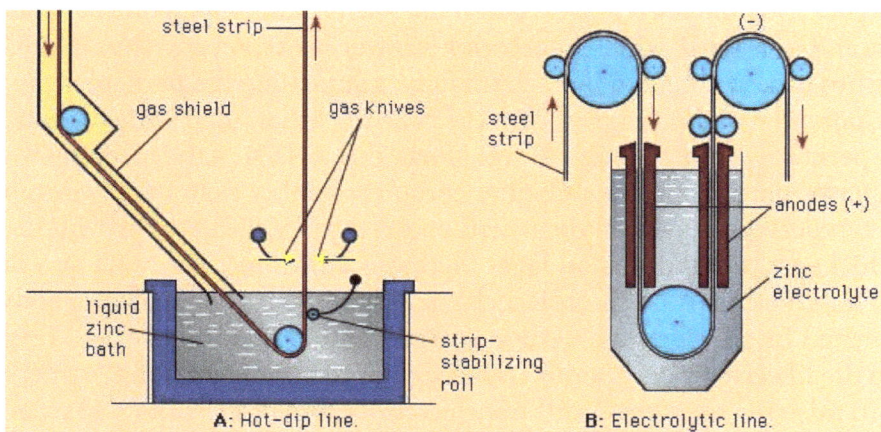

Principles of (A) hot-dip and (B) electrolytic galvanizing.

Electrodeposits of cadmium are also used for galvanic protection of steel. Hot-dip aluminum-coated steel is used in the exhaust systems of automobiles. At low temperatures its action is primarily galvanic, while at high temperatures it oxidizes to form a barrier layer.

Protective films on steel are susceptible to being broken by flying stones or other sharp objects. This is especially true on the parts of an automobile that are close to the road. The barriers described above have limited ability for self-healing after they are broken, but protection of the underlying small exposed area of steel can be maintained if a metal that has a greater tendency than steel to give up electrons in water is attached to the surface. In this way, when the protective barrier is broken, the more reactive metal is corroded preferentially, and the steel is given "galvanic" protection.

A layer of zinc can be placed on a steel surface either by hot-dipping the steel in molten zinc or by electroplating zinc onto the surface. Galvanizedsteel is much more resistant to corrosion than ungalvanized steel. For this reason, it is often used in the lowest panels of automobiles—*i.e.,* the parts exposed to corrosive salt spray from snowy roads.

Other Coating Techniques

Among other methods for applying metal layers to metal is thermal spray coating, a generic term for processes in which a metal wire is melted by a plasma arc or a flame, atomized, and sprayed onto a surface in an inert gas. Another method is vacuum coating, which produces thin, bright, attractive layers on a part by evaporating and depositing a coating metal in a high vacuum.

Enhanced oxidation protection is imparted to high-temperature turbine parts made of superalloys by annealing the parts in a container containing volatile aluminum chloride. This is done at high temperatures, so that aluminum diffuses into the alloy to form an aluminum-rich surface layer. In high-temperature service, such a layer oxidizes to form a protective aluminum oxide coating.

Hardening

Carburizing

The strength of hardened steel increases rapidly as the percentage of carbon is increased, but at the same time the steel's toughness decreases. Often the most useful part is one in which the surface is higher in carbon and thus hard, while the interior is lower in carbon and thus tough. Such a combination of properties can be obtained by carburizing, or annealing the parts in a gas rich in carbon. The carburizing potential of the gas rises with the ratio of carbon monoxide to carbon dioxide. The carburizing temperature is high enough to transform the surface of the steel to the high-temperature austenite phase, which has a much higher carbon solubility than the low-temperature ferrite phase. At these temperatures, carbon deposited on the surface diffuses through the steel and into the solid. The thickness of the diffusion layer increases with time, although at a decreasing rate; depths of 1 to 2 millimetres (0.04 to 0.08 inch) in 4 to 16 hours are typical. Following diffusion, the part is quenched in oil. The high-carbon surface transforms into a hard, brittle martensitic structure, while the lower-carbon interior transforms into a tougher, softer structure. The part is then tempered to raise the toughness of the surface layer. Small machine parts, such as gears, are often carburized to increase their strength and resistance to wear.

Nitriding

Nitriding provides an alternative means of hardening a steel surface. The surface layer is only one-tenth the depth of a carburized layer, but it is appreciably harder. The steel part is heated to a lower temperature, so that its crystal structure remains ferritic. Heating is conducted in an atmosphere of ammonia (NH_3) and hydrogen, and nitrogen from the ammonia diffuses into the steel.

Hardening is accomplished in one of two ways. One way is solid-solution hardening, which occurs in all steels. The other way is precipitation hardening. For example, if a steel contains aluminum, the aluminum and nitrogen will combine to form very fine particles that harden the steel quite effectively.

Though it is extremely hard, the nitride layer does not tend to crack, because it is very thin and adheres well to the ductile steel beneath it. The part need not be quenched from the nitriding temperature, nor does it need to be tempered before it is put into service.

Other Methods

Surfaces can also be hardened by heat treating with induction or laser heating. In other applications, surface hardening is accomplished by depositing a "hard-facing alloy" on the base part. One example is "hard chromium plating," in which a thick layer of chromium is deposited on a part. Automotive valve stems and piston rings and the bores of diesel engine cylinders are common applications.

Metallography and Metal Testing

Metallography

The properties of an alloy of a given composition can change markedly with the microscopic arrangement of its crystalline grains—i.e., its microstructure. To evaluate and control the microstructure of a sample, various types of microscope are used, and the field is called metallography.

Optical Microscopy

The simplest, and oldest, type of metallography (though hardly a century old) involves polishing the surface to a mirrorlike finish and examining light reflected from it at magnifications of 50 to 1500×. If the surface is lightly etched in an appropriate solution (often an acid), the grain boundaries, matrix, and constituent phases will be attacked at different rates and will be discernible. This makes it possible to establish which phases are present as well as their shape, size, and distribution. Similarly, grain size and shape can be observed. With this information it is possible to infer the history of the sample and predict its behaviour. Metallography is of particular value in the analysis of samples that have failed or performed in an unexpected manner.

Electron Microscopy

Great progress has been made in using finely focused beams of energetic electrons to examine metals. Electron microscopes are basically of two types, transmission and scanning. Transmission electron microscopes require the preparation of films so thin that they are transparent to a beam of electrons with energies of roughly 200 kiloelectron volts. This means the film must have a thickness of only one, or a few, hundred nanometres (10^{-9}metre). Films of lighter elements, such as aluminum, can be thicker, while films of heavier elements, such as gold, must be thinner. Contrast between neighbouring regions is best developed by differences in their diffraction of the electron beam, although differences in density can also be used. Spatial resolution is excellent, going down to atomic resolution in special microscopes, and orientation relations between adjoining regions can be easily discerned. On the other hand, only very small samples can be examined in any given film. This means that the technique is not good for measuring defects larger than the film thickness or those whose number per unit volume is low.

A scanning electron microscope (SEM) uses a narrow beam of electrons (often of about 40 kiloelectron volts) that scans the surface of a sample and forms a corresponding image from the backscattered electrons or secondary electrons. No special surface preparation is necessary, and, since the depth of focus in an SEM is much greater than in an optical microscope, quite irregular surfaces, such as fractures, can be studied successfully. Useful magnifications range from 100 to 20,000×.

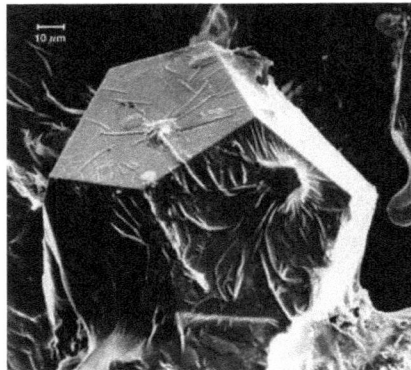

A scanning electron microscope (SEM) image of quasicrystalline
aluminum-copper-iron, revealing the pentagonal dodecahedral shape of the grain.

The electron beam used in an SEM causes each atom near the surface to emit an X ray that is characteristic of that element. By constructing an image based on the distribution of the intensity of the characteristic X ray of a given element, it is possible to show that element's distribution among the phases in the surface. If the electron beam is not swept but held in one spot, a chemical analysis can be made of the various elements in the region under the electron beam by measuring the intensity of the X rays emitted by each element.

Testing Mechanical Properties

The most common mechanical properties are yield stress, elongation, hardness, and toughness. The first two are measured in a tensile test, where a sample is loaded until it begins to undergo plastic strain (*i.e.,* strain that is not recovered when the sample is unloaded). This stress is called the yield stress. It is a property that is the same for various samples of the same alloy, and it is useful in designing structures since it predicts the loads beyond which a structure will be permanently bent out of shape.

If the tensile test is continued past yielding, the load reaches a maximum as the strain localizes and a neck develops in the sample. The maximum load, divided by the initial cross-sectional area of the sample, is called the ultimate tensile stress (UTS). The final length minus the initial length, divided by the initial length, is called the elongation. Yield stress, UTS, and elongation are the most commonly tabulated mechanical properties of metals.

The hardness of a metal can be measured in several ways. If a hard indenter (a sphere, cone, or pyramid) is pushed a short distance into a metal with a defined load, the load divided by the contact area becomes the measure of hardness. For testing steel, one of the oldest of such tests, the Brinell hardness test, uses a 10-millimetre-diameter ball and a 3,000-kilogram load. Brinell hardness values correlate well with UTS. Much smaller loads and diamond microindenters also can be used in conjunction with a microscope to measure the hardness of quite small regions (down to a few micrometres, or millionths of a metre, across).

A different type of testing machine, which indicates hardness directly, bears the name Rockwell. Here, instead of measuring the width of the indentation after the indenter has been removed, a sensitive gauge indicates the depth to which the indenter sinks into the surface under a given load. Various sizes of indenters and loads allow a wide range of hardness to be measured. These hardness numbers are useful for quality control in manufacturing, especially in assuring consistency from batch to batch.

Of particular concern in engineering structures is the prevention of sudden, complete failure, as in the fracture of a brittle material. Much to be preferred is a structure that will deform under an overload but not fail. Sudden failure begins at a notch or crack that locally concentrates the stress, and the energy required to extend such a crack in a solid is a measure of the solid's toughness. In a hard, brittle material, toughness is low, while in a strong, ductile metal it is high. A common test of toughness is the Charpy test, which employs a small bar of a metal with a V-shaped groove cut on one side. A large hammer is swung so as to strike the bar on the side opposite the groove. The energy absorbed in driving the hammer through the bar is the toughness.

A test requiring more instrumentation, but measuring material properties that are more useful for analysis, is the pulling apart of two sides of a sample containing a crack that is initially cut about one-third of the way through the sample. The use and analysis of such a test is called fracture mechanics, and the information acquired is used to demonstrate the integrity of structures made of strong materials that contain small flaws—for example, rocket casings, airplanes, and nuclear reactor pressure vessels.

If a part is loaded once to a stress near the yield stress, it will not break. However, if it is loaded repeatedly to this level, it will eventually break. This failure is called fatigue, and avoiding fatigue is an important goal in the design of moving machinery. The more cycles a part will undergo, the lower the allowable stress it must be assigned in order to avoid failure by fatigue.

Another type of failure observable at loads below the yield stress is called creep. If a load is applied and left on the sample for months or years, the sample will slowly extend. In metals with high melting temperatures, creep becomes a problem at higher temperatures. It becomes a limiting consideration in gas turbines that operate at the highest temperature the metal parts can take.

References

- Meyers, M. A.; Chen, P. Y.; Lin, A. Y. M.; Seki, Y. (2008). "Biological materials: Structure and mechanical properties". Progress in Materials Science. 53: 1–206. Doi:10.1016/j.pmatsci.2007.05.002

- What-is-ceramic-engineering-scope-and-career-opportunities, unique-courses, courses: careerindia.com, Retrieved 1 January, 2019

- "Polypropylene (PP) Plastic: Types, Properties, Uses & Structure Info". Omnexus.specialchem.com. Retrieved 2019-03-17

- Snigirev, A. (2007). "Two-step hard X-ray focusing combining Fresnel zone plate and single-bounce ellipsoidal capillary". Journal of Synchrotron Radiation. 14 (Pt 4): 326–330. Doi:10.1107/S0909049507025174. PMID 17587657

- Metallurgy, science: britannica.com, Retrieved 2 February, 2019

- "Bakelite: The World's First Synthetic Plastic". National Historic Chemical Landmarks. American Chemical Society. Archived from the original on July 22, 2012. Retrieved June 25, 2012

- Hill, Joanne; Mulholland, Gregory; Persson, Kristin; Seshadri, Ram; Wolverton, Chris; Meredig, Bryce (4 May 2016). "Materials science with large-scale data and informatics: Unlocking new opportunities". MRS Bulletin. 41 (5): 399–409. Doi:10.1557/mrs.2016.93

3

Semiconductors: Principles and Concepts

A material whose electrical conductivity value lies between that of a conductor and an insulator is referred to as a semiconductor. The major types of semiconductors are intrinsic semiconductors and extrinsic semiconductors. This chapter has been carefully written to provide an easy understanding of these types of semiconductors.

A semiconductor is a solid whose electrical conductivity can be controlled over a wide range, either permanently or dynamically. Semiconductors are tremendously important technologically and economically. Semiconductors are essential materials in all modern electrical devices, from computers to cellular phones to digital audio players. Silicon is the most commercially important semiconductor, though dozens of others are important as well.

Semiconductor devices are electronic components that exploit the electronic properties of semiconductor materials, principally silicon, germanium, and gallium arsenide. Semiconductor devices have replaced thermionic devices (vacuum tubes) in most applications. They use electronic conduction in the solid state as opposed to the gaseous state or thermionic emission in a high vacuum.

Semiconductor devices are manufactured as single, discrete devices or integrated circuits (ICs), which consist of a number—from a few devices to millions—of devices manufactured onto a single semiconductor substrate.

Semiconductors are very similar to insulators. The two categories of solids differ primarily in that insulators have larger band gaps—energies that electrons must acquire to be free to flow. In semiconductors at room temperature, just as in insulators, very few electrons gain enough thermal energy to leap the band gap, which is necessary for conduction. For this reason, pure semiconductors and insulators, in the absence of applied fields, have roughly similar electrical properties. The smaller bandgaps of semiconductors, however, allow for many other means besides temperature to control their electrical properties.

Semiconductors' intrinsic electrical properties are very often permanently modified by introducing impurities, in a process known as doping. Usually it is reasonable to approximate that each impurity atom adds one electron or one "hole" that may flow freely. Upon the addition of a sufficiently large proportion of dopants, semiconductors conduct electricitynearly as well as metals. The junctions between regions of semiconductors that are doped with different impurities contain built-in electric fields, which are critical to semiconductor device operation.

In addition to permanent modification through doping, the electrical properties of semiconductors are often dynamically modified by applying electric fields. The ability to control conductivity in small and well-defined regions of semiconductor material, statically through doping and dynamically through the application of electric fields, has led to the development of a broad array of semiconductor devices, like transistors. Semiconductor devices with dynamically controlled conductivity are the building blocks of integrated circuits, like the microprocessor. These "active" semiconductor devices are combined with simpler passive components, such as semiconductor capacitors and resistors, to produce a variety of electronic devices.

In certain semiconductors, when electrons fall from the conduction band to the valence band (the energy levels above and below the band gap), they often emit light. This photoemission process underlies the light-emitting diode (LED) and the semiconductor laser, both of which are tremendously important commercially. Conversely, semiconductor absorption of light in photodetectors excites electrons from the valence band to the conduction band, facilitating reception of fiber optic communications, and providing the basis for energy from solar cells.

Semiconductors may be elemental materials, such as silicon, compound semiconductors such as gallium arsenide, or alloys, such as silicon germanium or aluminium gallium arsenide.

Semiconductor Device Materials

By far, silicon (Si) is the most widely used material in semiconductor devices. Its combination of low raw material cost, relatively simple processing, and a useful temperature range make it currently the best compromise among the various competing materials. Silicon used in semiconductor device manufacturing is currently fabricated into boules that are large enough in diameter to allow the production of 300 mm (12 in.) wafers.

Germanium (Ge) was a widely used early semiconductor material but its thermal sensitivity makes it less useful than silicon. Today, germanium is often alloyed with silicon for use in very-high-speed SiGe devices; IBM is a major producer of such devices.

Gallium arsenide (GaAs) is also widely used in high-speed devices but so far, it has been difficult to form large-diameter boules of this material, limiting the wafer diameter to sizes significantly smaller than silicon wafers thus making mass production of GaAs devices significantly more expensive than silicon.

Other less common materials are also in use or under investigation.

Silicon carbide (SiC) has found some application as the raw material for blue light-emitting diodes (LEDs) and is being investigated for use in semiconductor devices that could withstand very high operating temperatures and environments with the presence of significant levels of ionizing radiation. IMPATT diodes have also been fabricated from SiC.

Various indium compounds (indium arsenide, indium antimonide, and indium phosphide) are also being used in LEDs and solid state laser diodes. Selenium sulfide is being studied in the manufacture of photovoltaic solar cells.

Preparation of Semiconductor Materials

Semiconductors with predictable, reliable electronic properties are necessary for mass production.

The level of chemical purity needed is extremely high because the presence of impurities even in very small proportions can have large effects on the properties of the material. A high degree of crystalline perfection is also required, since faults in crystal structure (such as dislocations, twins, and stacking faults) interfere with the semiconducting properties of the material. Crystalline faults are a major cause of defective semiconductor devices. The larger the crystal, the more difficult it is to achieve the necessary perfection. Current mass production processes use crystal ingots between four and twelve inches (300 mm) in diameter which are grown as cylinders and sliced into wafers.

Because of the required level of chemical purity, and the perfection of the crystal structure which are needed to make semiconductor devices, special methods have been developed to produce the initial semiconductor material. A technique for achieving high purity includes growing the crystal using the Czochralski process. An additional step that can be used to further increase purity is known as zone refining. In zone refining, part of a solid crystal is melted. The impurities tend to concentrate in the melted region, while the desired material recrystalizes leaving the solid material more pure and with fewer crystalline faults.

In manufacturing semiconductor devices involving heterojunctions between different semiconductor materials, the lattice constant, which is the length of the repeating element of the crystal structure, is important for determining the compatibility of materials.

Semiconductor Device Fundamentals

The main reason semiconductor materials are so useful is that the behaviour of a semiconductor can be easily manipulated by the addition of impurities, known as doping. Semiconductor conductivity can be controlled by introduction of an electric field, by exposure to light, and even pressure and heat; thus, semiconductors can make excellent sensors. Current conduction in a semiconductor occurs via mobile or "free" electrons and holes (collectively known as charge carriers). Doping a semiconductor such as silicon with a small amount of impurity atoms, such as phosphorus or boron, greatly increases the number of free electrons or holes within the semiconductor. When a doped semiconductor contains excess holes it is called "p-type," and when it contains excess free electrons it is known as "n-type." The semiconductor material used in devices is doped under highly controlled conditions in a fabrication facility, or fab, to precisely control the location and concentration of p- and n-type dopants. The junctions which form where n-type and p-type semiconductors join together are called p-n junctions.

Diode

The p-n junction diode is a device made from a p-n junction. At the junction of a p-type and an n-type semiconductor there forms a region called the depletion zone which blocks current conduction from the n-type region to the p-type region, but allows current to conduct from the p-type region to the n-type region. Thus when the device is forward biased, with the p-side at higher electric potential, the diode conducts current easily; but the current is very small when the diode is reverse biased.

Exposing a semiconductor to light can generate electron–hole pairs, which increases the number of free carriers and its conductivity. Diodes optimized to take advantage of this phenomenon are known as photodiodes. Compound semiconductor diodes can also be used to generate light, as in light-emitting diodes and laser diodes.

Transistor

Bipolar junction transistors are formed from two p-n junctions, in either n-p-n or p-n-p configuration. The middle, or base, region between the junctions is typically very narrow. The other regions, and their associated terminals, are known as the emitter and the collector. A small current injected through the junction between the base and the emitter changes the properties of the base-collector junction so that it can conduct current even though it is reverse biased. This creates a much larger current between the collector and emitter, controlled by the base-emitter current.

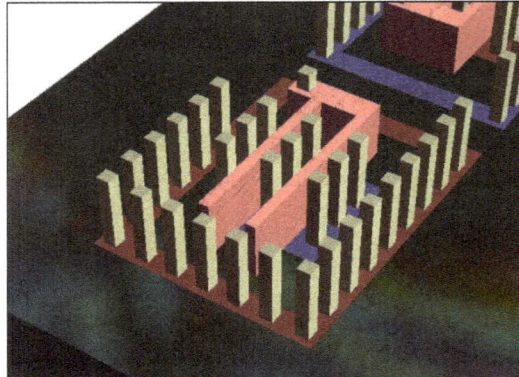

Two MOS transistors with common gate
(metallic layers and dielectric removed for clarity).

Another type of transistor, the field effect transistor operates on the principle that semiconductor conductivity can be increased or decreased by the presence of an electric field. An electric field can increase the number of free electrons and holes in a semiconductor, thereby changing its conductivity. The field may be applied by a reverse-biased p-n junction, forming a junction field effect transistor, or JFET; or by an electrode isolated from the bulk material by an oxide layer, forming a metal-oxide-semiconductor field effect transistor, or MOSFET.

The MOSFET is the most used semiconductor device today. The gate electrode is charged to produce an electric field that controls the conductivity of a "channel" between two terminals, called the source and drain. Depending on the type of carrier in the channel, the device may be an n-channel (for electrons) or a p-channel (for holes) MOSFET. Although the MOSFET is named in part for its "metal" gate, in modern devices polysilicon is typically used instead.

Cross-section through a MOS transistor (metallic layers
and dielectric removed for clarity), foreground.

Semiconductor Device Applications

All transistor types can be used as the building blocks of logic gates, which are fundamental in the design of digital circuits. In digital circuits like microprocessors, transistors act as on-off switches; in the MOSFET, for instance, the voltage applied to the gate determines whether the switch is on or off.

Transistors used for analog circuits do not act as on-off switches; rather, they respond to a continuous range of inputs with a continuous range of outputs. Common analog circuits include amplifiers and oscillators.

Circuits that interface or translate between digital circuits and analog circuits are known as mixed-signal circuits.

Power semiconductor devices are discrete devices or integrated circuits intended for high current or high voltage applications. Power integrated circuits combine IC technology with power semiconductor technology, these are sometimes referred to as "smart" power devices. Several companies specialize in manufacturing power semiconductors.

Component Identifiers

The type designators of semiconductor devices are often manufacturer specific. Nevertheless, there have been attempts at creating standards for type codes, and a subset of devices follow those. For discrete devices, for example, there are three standards: JEDEC JESD370B in USA, Pro Electron in Europe and JIS in Japan.

Physics of Semiconductors

Band Structure

Like other solids, the electrons in semiconductors can have energies only within certain bands between the energy of the ground state, corresponding to electrons tightly bound to the atomic nuclei of the material, and the free electron energy, which is the energy required for an electron to escape entirely from the material. The energy bands each correspond to a large number of discrete quantum states of the electrons, and most of the states with low energy are full, up to a particular band called the valence band. Semiconductors and insulators are distinguished from metals because the valence band in the former materials is very nearly full under normal conditions.

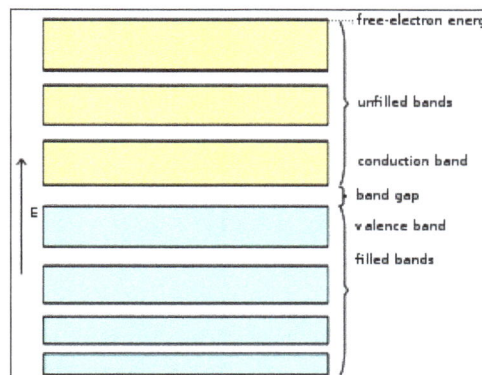

Band structure of a semiconductor showing a full valence band and an empty conduction band.

The ease with which electrons in a semiconductor can be excited from the valence band to the conduction band depends on the band gap between the bands, and it is the size of this energy bandgap that serves as an arbitrary dividing line (roughly 4 eV) between semiconductors and insulators.

The electrons must move between states to conduct electric current, and so due to the Pauli exclusion principle full bands do not contribute to the electrical conductivity. However, as the temperature of a semiconductor rises above absolute zero, the states of the electrons are increasingly randomized, or smeared out, and some electrons are likely to be found in states of the *conduction band*, which is the band immediately above the valence band. The current-carrying electrons in the conduction band are known as "free electrons," although they are often simply called "electrons" if context allows this usage to be clear.

Electrons excited to the conduction band also leave behind electron holes, or unoccupied states in the valence band. Both the conduction band electrons and the valence band holes contribute to electrical conductivity. The holes themselves don't actually move, but a neighboring electron can move to fill the hole, leaving a hole at the place it has just come from, and in this way the holes appear to move, and the holes behave as if they were actual positively charged particles.

This behavior may also be viewed in relation to chemical bonding. The electrons that have enough energy to be in the conduction band have broken free of the covalent bonds between neighboring atoms in the solid, and are free to move around, and hence conduct charge.

It is an important distinction between conductors and semiconductors that, in semiconductors, movement of charge (current) is facilitated by both electrons and holes. Contrast this to a conductor where the Fermi level lies *within* the conduction band, such that the band is only half filled with electrons. In this case, only a small amount of energy is needed for the electrons to find other unoccupied states to move into, and hence for current to flow.

Carrier Generation and Recombination

When ionizing radiation strikes a semiconductor, it may excite an electron out of its energy level and consequently leave a hole. This process is known as *electron–hole pair generation*. Electron-hole pairs are constantly generated from thermal energy as well, in the absence of any external energy source.

Electron-hole pairs are also apt to recombine. Conservation of energy demands that these recombination events, in which an electron loses an amount of energy larger than the band gap, be accompanied by the emission of thermal energy (in the form of phonons) or radiation (in the form of photons).

INTRINSIC SEMICONDUCTOR

An intrinsic (pure) semiconductor, also called an undoped semiconductor or i-type semiconductor, is a pure semiconductor without any significant dopant species present. The number of charge carriers is therefore determined by the properties of the material itself instead of the amount of impurities. In intrinsic semiconductors the number of excited electrons and the number of holes are equal: n = p. This may even be the case after doping the semiconductor, though only if it is doped with both donors and acceptors equally. In this case, n = p still holds, and the semiconductor remains intrinsic, though doped.

The electrical conductivity of intrinsic semiconductors can be due to crystallographic defects or electron excitation. In an intrinsic semiconductor the number of electrons in the conduction band is equal to the number of holes in the valence band. An example is $Hg_{0.8}Cd_{0.2}Te$ at room temperature.

An indirect band gap intrinsic semiconductor is one in which the maximum energy of the valence band occurs at a different k (k-space wave vector) than the minimum energy of the conduction band. Examples include silicon and germanium. A direct band gap intrinsic semiconductor is one where the maximum energy of the valence band occurs at the same as the minimum energy of the conduction band. Examples include gallium arsenide.

A silicon crystal is different from an insulator because at any temperature above absolute zero, there is a non-zero probability that an electron in the lattice will be knocked loose from its position, leaving behind an electron deficiency called a "hole". If a voltage is applied, then both the electron and the hole can contribute to a small current flow.

The conductivity of a semiconductor can be modeled in terms of the band theory of solids. The band model of a semiconductor suggests that at ordinary temperatures there is a finite possibility that electrons can reach the conduction band and contribute to electrical conduction.

The term intrinsic here distinguishes between the properties of pure "intrinsic" silicon and the dramatically different properties of doped n-type or p-type semiconductors.

Electrons and Holes

In an intrinsic semiconductor such as silicon at temperatures above absolute zero, there will be some electrons which are excited across the band gap into the conduction band and which can support charge flowing. When the electron in pure silicon crosses the gap, it leaves behind an electron vacancy or "hole" in the regular silicon lattice. Under the influence of an external voltage, both the electron and the hole can move across the material. In an n-type semiconductor, the dopant contributes extra electrons, dramatically increasing the conductivity. In a p-type semiconductor, the dopant produces extra vacancies or holes, which likewise increase the conductivity. It is however the behavior of the p-n junction which is the key to the enormous variety of solid-state electronic devices.

Semiconductor Current

The current which will flow in an intrinsic semiconductor consists of both electron and hole current. That is, the electrons which have been freed from their lattice positions into the conduction band can move through the material. In addition, other electrons can hop between lattice positions to fill the vacancies left by the freed electrons. This additional mechanism is called hole conduction because it is as if the holes are migrating across the material in the direction opposite to the free electron movement. The current flow in an intrinsic semiconductor is influenced by the density of energy states which in turn influences the electron density in the conduction band. This current is highly temperature dependent.

EXTRINSIC SEMICONDUCTOR

An extrinsic semiconductor is one that has been doped; during manufacture of the semiconductor crystal a trace element or chemical called a doping agent has been incorporated chemically into the crystal, for the purpose of giving it different electrical properties than the pure semiconductor crystal, which is called an intrinsic semiconductor. In an extrinsic semiconductor it is these foreign dopant atoms in the crystal lattice that mainly provide the charge carriers which carry electric current through the crystal. The doping agents used are of two types, resulting in two types of extrinsic semiconductor. An electron donor dopant is an atom which, when incorporated in the crystal, releases a mobile conduction electron into the crystal lattice. An extrinsic semiconductor which has been doped with electron donor atoms is called an n-type semiconductor, because the majority of charge carriers in the crystal are negative electrons. An electron acceptor dopant is an atom which accepts an electron from the lattice, creating a vacancy where an electron should be called a hole which can move through the crystal like a positively charged particle. An extrinsic semiconductor which has been doped with electron acceptor atoms is called a p-type semiconductor, because the majority of charge carriers in the crystal are positive holes.

Doping is the key to the extraordinarily wide range of electrical behavior that semiconductors can exhibit, and extrinsic semiconductors are used to make semiconductor electronic devices such as diodes, transistors, integrated circuits, semiconductor lasers, LEDs, and photovoltaic cells. Sophisticated semiconductor fabrication processes like photolithography can implant different dopant elements in different regions of the same semiconductor crystal wafer, creating semiconductor devices on the wafer's surface. For example a common type of transistor, the n-p-n bipolar transistor, consists of an extrinsic semiconductor crystal with two regions of n-type semiconductor, separated by a region of p-type semiconductor, with metal contacts attached to each part.

Conduction in Semiconductors

A solid substance can conduct electric current only if it contains charged particles, electrons, which are free to move about and not attached to atoms. In a metal conductor, it is the metal atoms that provide the electrons; typically each metal atom releases one of its outer orbital electrons to become a conduction electron which can move about throughout the crystal, and carry electric current. Therefore the number of conduction electrons in a metal is equal to the number of atoms, a very large number, making metals good conductors.

Unlike in metals, the atoms that make up the bulk semiconductor crystal do not provide the electrons which are responsible for conduction. In semiconductors, electrical conduction is due to the mobile charge carriers, electrons or holes which are provided by impurities or dopant atoms in the crystal. In an extrinsic semiconductor, the concentration of doping atoms in the crystal largely determines the density of charge carriers, which determines its electrical conductivity, as well as a great many other electrical properties. This is the key to semiconductors' versatility; their conductivity can be manipulated over many orders of magnitude by doping.

Semiconductor Doping

Semiconductor doping is the process that changes an intrinsic semiconductor to an extrinsic semiconductor. During doping, impurity atoms are introduced to an intrinsic semiconductor. Impurity atoms are atoms of a different element than the atoms of the intrinsic semiconductor. Impurity atoms act as either donors or acceptors to the intrinsic semiconductor, changing the electron and hole concentrations of the semiconductor. Impurity atoms are classified as either donor or acceptor atoms based on the effect they have on the intrinsic semiconductor.

Donor impurity atoms have more valence electrons than the atoms they replace in the intrinsic semiconductor lattice. Donor impurities "donate" their extra valence electrons to a semiconductor's conduction band, providing excess electrons to the intrinsic semiconductor. Excess electrons increase the electron carrier concentration (n_o) of the semiconductor, making it n-type.

Acceptor impurity atoms have fewer valence electrons than the atoms they replace in the intrinsic semiconductor lattice. They "accept" electrons from the semiconductor's valence band. This provides excess holes to the intrinsic semiconductor. Excess holes increase the hole carrier concentration (p_o) of the semiconductor, creating a p-type semiconductor.

Semiconductors and dopant atoms are defined by the column of the periodic table in which they fall. The column definition of the semiconductor determines how many valence electrons its atoms have and whether dopant atoms act as the semiconductor's donors or acceptors.

Group IV semiconductors use group V atoms as donors and group III atoms as acceptors.

Group III-V semiconductors, the compound semiconductors, use group VI atoms as donors and group II atoms as acceptors. Group III-V semiconductors can also use group IV atoms as either donors or acceptors. When a group IV atom replaces the group III element in the semiconductor lattice, the group IV atom acts as a donor. Conversely, when a group IV atom replaces the group V element, the group IV atom acts as an acceptor. Group IV atoms can act as both donors and acceptors; therefore, they are known as amphoteric impurities.

	Intrinsic semiconductor	Donor atoms	Acceptor atoms
Group IV semiconductors	Silicon, Germanium	Phosphorus, Arsenic, Antimony	Boron, Aluminium, Gallium
Group III-V semiconductors	Aluminum phosphide, Aluminum arsenide, Gallium arsenide, Gallium nitride	Selenium, Tellurium, Silicon, Germanium	Beryllium, Zinc, Cadmium, Silicon, Germanium

The Two Types of Semiconductor

N-type Semiconductors

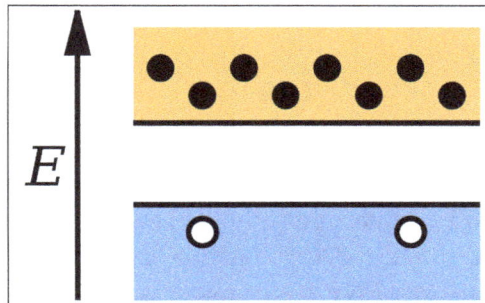

Band structure of an n-type semiconductor.

Dark circles in the conduction band are electrons and light circles in the valence band are holes. The image shows that the electrons are the majority charge carrier.

N-type semiconductors are created by doping an intrinsic semiconductor with an electron donor element during manufacture. The term *n-type* comes from the negative charge of the electron. In *n-type* semiconductors, electrons are the majority carriers and holes are the minority carriers. A common dopant for *n-type* silicon is phosphorus or arsenic. In an *n-type* semiconductor, the Fermi level is greater than that of the intrinsic semiconductor and lies closer to the conduction band than the valence band.

P-type Semiconductors

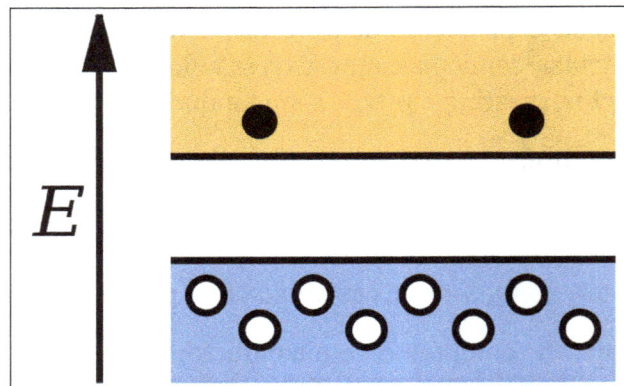

Band structure of a p-type semiconductor.

Dark circles in the conduction band are electrons and light circles in the valence band are holes. The image shows that the holes are the majority charge carrier

P-type semiconductors are created by doping an intrinsic semiconductor with an electron acceptor element during manufacture. The term p-type refers to the positive charge of a hole. As opposed to n-type semiconductors, p-type semiconductors have a larger hole concentration than electron concentration. In p-type semiconductors, holes are the majority carriers and electrons are the minority carriers. A common p-type dopant for silicon is boron or gallium. For p-type semiconductors the Fermi level is below the intrinsic Fermi level and lies closer to the valence band than the conduction band.

Use of Extrinsic Semiconductors

Extrinsic semiconductors are components of many common electrical devices. A semiconductor diode (devices that allow current in only one direction) consists of p-type and n-type semiconductors placed in junction with one another. Currently, most semiconductor diodes use doped silicon or germanium.

Transistors (devices that enable current switching) also make use of extrinsic semiconductors. Bipolar junction transistors (BJT), which amplify current, are one type of transistor. The most common BJTs are NPN and PNP type. NPN transistors have two layers of n-type semiconductors sandwiching a p-type semiconductor. PNP transistors have two layers of p-type semiconductors sandwiching an n-type semiconductor.

Field-effect transistors (FET) are another type of transistor which amplify current implementing extrinsic semiconductors. As opposed to BJTs, they are called unipolar because they involve single carrier type operation – either N-channel or P-channel. FETs are broken into two families, junction gate FET (JFET), which are three terminal semiconductors, and insulated gate FET (IGFET), which are four terminal semiconductors.

Other devices implementing the extrinsic semiconductor:

- Lasers

- Solar cells

- Photodetectors

- Light-emitting diodes

- Thyristors

DOPING

In semiconductor production, doping is the intentional introduction of impurities into an intrinsic semiconductor for the purpose of modulating its electrical, optical and structural properties. The doped material is referred to as an extrinsic semiconductor. A semiconductor doped to such high levels that it acts more like a conductor than a semiconductor is referred to as a degenerate semiconductor.

In the context of phosphors and scintillators, doping is better known as activation. Doping is also used to control the color in some pigments.

Carrier Concentration

The concentration of the dopant used affects many electrical properties. Most important is the material's charge carrier concentration. In an intrinsic semiconductor under thermal equilibrium, the concentrations of electrons and holes are equivalent. That is,

$$n = p = n_i.$$

In a non-intrinsic semiconductor under thermal equilibrium, the relation becomes (for low doping):

$$n_0 \cdot p_0 = n_i^2$$

where n_o is the concentration of conducting electrons, p_o is the electron hole concentration, and n_i is the material's intrinsic carrier concentration. The intrinsic carrier concentration varies between materials and is dependent on temperature. Silicon's n_i, for example, is roughly 1.08×10^{10} cm^{-3} at 300 kelvins, about room temperature.

In general, increased doping leads to increased conductivity due to the higher concentration of carriers. Degenerate (very highly doped) semiconductors have conductivity levels comparable to metals and are often used in integrated circuits as a replacement for metal. Often superscript plus and minus symbols are used to denote relative doping concentration in semiconductors. For example, n^+ denotes an n-type semiconductor with a high, often degenerate, doping concentration. Similarly, p^- would indicate a very lightly doped p-type material. Even degenerate levels of doping imply low concentrations of impurities with respect to the base semiconductor. In intrinsic crystalline silicon, there are approximately 5×10^{22} atoms/cm³. Doping concentration for silicon semiconductors may range anywhere from 10^{13} cm⁻³ to 10^{18} cm⁻³. Doping concentration above about 10^{18} cm⁻³ is considered degenerate at room temperature. Degenerately doped silicon contains a proportion of impurity to silicon on the order of parts per thousand. This proportion may be reduced to parts per billion in very lightly doped silicon. Typical concentration values fall somewhere in this range and are tailored to produce the desired properties in the device that the semiconductor is intended for.

Effect on Band Structure

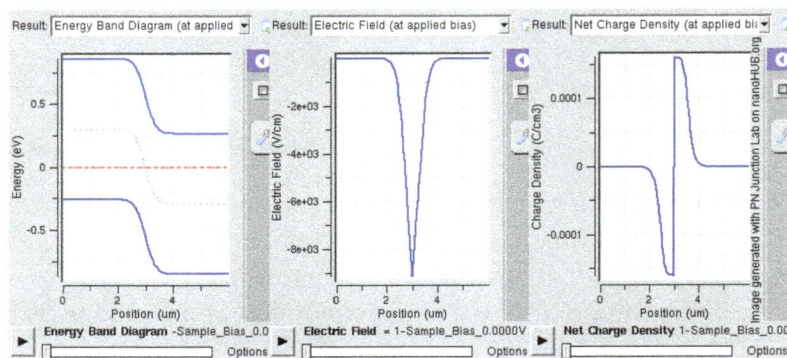

Band diagram of PN junction operation in forward bias mode showing reducing depletion width. Both p and n junctions are doped at a 1×10^{15}/cm³ doping level, leading to built-in potential of ~0.59 V. Reducing depletion width can be inferred from the shrinking charge profile, as fewer dopants are exposed with increasing forward bias.

Doping a semiconductor in a good crystal introduces allowed energy states within the band gap, but very close to the energy band that corresponds to the dopant type. In other words, electron donor impurities create states near the conduction band while electron acceptor impurities create states near the valence band. The gap between these energy states and the nearest energy band is usually referred to as dopant-site bonding energy or E_B and is relatively small. For example, the E_B for boron in silicon bulk is 0.045 eV, compared with silicon's band gap of about 1.12 eV. Because E_B is so small, room temperature is hot enough to thermally ionize practically all of the dopant atoms and create free charge carriers in the conduction or valence bands.

Dopants also have the important effect of shifting the energy bands relative to the Fermi level. The energy band that corresponds with the dopant with the greatest concentration ends up closer to the Fermi level. Since the Fermi level must remain constant in a system in thermodynamic equilibrium, stacking layers of materials with different properties leads to many useful electrical properties induced by band bending, if the interfaces can be made cleanly enough. For example, the p-n junction's properties are due to the band bending that happens as a result of the necessity to line up the bands in contacting regions of p-type and n-type material. This effect is shown in a band diagram. The band diagram typically indicates the variation in the valence band and conduction band edges versus some spatial dimension, often denoted x. The Fermi level is also usually indicated in the diagram. Sometimes the *intrinsic Fermi level*, E_i, which is the Fermi level in the absence of doping, is shown. These diagrams are useful in explaining the operation of many kinds of semiconductor devices.

Relationship to Carrier Concentration (Low Doping)

For low levels of doping, the relevant energy states are populated sparsely by electrons (conduction band) or holes (valence band). It is possible to write simple expressions for the electron and hole carrier concentrations, by ignoring Pauli exclusion (via Maxwell–Boltzmann statistics):

$$n_e = N_C(T)\exp((E_F - E_C)/kT), \quad n_h = N_V(T)\exp((E_V - E_F)/kT),$$

where E_F is the Fermi level, E_C is the minimum energy of the conduction band, and E_V is the maximum energy of the valence band. These are related to the value of the intrinsic concentration via

$$n_i^2 = n_h n_e = N_V(T)N_C(T)\exp((E_V - E_C)/kT),$$

an expression which is independent of the doping level, since $E_C - E_V$ (the band gap) does not change with doping.

The concentration factors $N_C(T)$ and $N_V(T)$ are given by

$$N_C(T) = 2(2\pi m_e^* kT / h^2)^{3/2} \quad N_V(T) = 2(2\pi m_h^* kT / h^2)^{3/2}.$$

where m_e^* and m_h^* are the density of states effective masses of electrons and holes, respectively, quantities that are roughly constant over temperature.

Techniques of Doping and Synthesis

The synthesis of n-type semiconductors may involve the use of vapor-phase epitaxy. In vapor-phase epitaxy, a gas containing the negative dopant is passed over the substrate wafer. In the case of n-type GaAs doping, hydrogen sulfide is passed over the gallium arsenide, and sulfur is incorporated into the structure. This process is characterized by a constant concentration of sulfur on the surface. In the case of semiconductors in general, only a very thin layer of the wafer needs to be doped in order to obtain the desired electronic properties. The reaction conditions typically range from 600 to 800 °C for the n-doping with group VI elements, and the time is typically 6–12 hours depending on the temperature.

Process

Some dopants are added as the (usually silicon) boule is grown, giving each wafer an almost uniform initial doping. To define circuit elements, selected areas — typically controlled by photolithography — are further doped by such processes as diffusion and ion implantation, the latter method being more popular in large production runs because of increased controllability.

Small numbers of dopant atoms can change the ability of a semiconductor to conduct electricity. When on the order of one dopant atom is added per 100 million atoms, the doping is said to be low or light. When many more dopant atoms are added, on the order of one per ten thousand atoms, the doping is referred to as high or heavy. This is often shown as n+ for n-type doping or p+ for p-type doping.

Dopant Elements

Group IV Semiconductors

For the Group IV semiconductors such as diamond, silicon, germanium, silicon carbide, and silicon germanium, the most common dopants are acceptors from Group III or donors from Group V elements. Boron, arsenic, phosphorus, and occasionally gallium are used to dope silicon. Boron is the p-type dopant of choice for silicon integrated circuit production because it diffuses at a rate that makes junction depths easily controllable. Phosphorus is typically used for bulk-doping of silicon wafers, while arsenic is used to diffuse junctions, because it diffuses more slowly than phosphorus and is thus more controllable.

By doping pure silicon with Group V elements such as phosphorus, extra valence electrons are added that become unbonded from individual atoms and allow the compound to be an electrically conductive n-type semiconductor. Doping with Group III elements, which are missing the fourth valence electron, creates "broken bonds" (holes) in the silicon lattice that are free to move. The result is an electrically conductive p-type semiconductor. In this context, a Group V element is said to behave as an electron donor, and a group III element as an acceptor. This is a key concept in the physics of a diode.

A very heavily doped semiconductor behaves more like a good conductor (metal) and thus exhibits more linear positive thermal coefficient. Such effect is used for instance in sensistors. Lower dosage of doping is used in other types (NTC or PTC) thermistors.

Silicon Dopants

- Acceptors, p-type
  - Boron is a p-type dopant. Its diffusion rate allows easy control of junction depths. Common in CMOS technology. Can be added by diffusion of diborane gas. The only acceptor with sufficient solubility for efficient emitters in transistors and other applications requiring extremely high dopant concentrations. Boron diffuses about as fast as phosphorus.
  - Aluminium, used for deep p-diffusions. Not popular in VLSI and ULSI. Also a common unintentional impurity.
  - Gallium is a dopant used for long-wavelength infrared photoconduction silicon detectors in the 8–14 μm atmospheric window. Gallium-doped silicon is also promising for solar cells, due to its long minority carrier lifetime with no lifetime degradation; as such it is gaining importance as a replacement of boron doped substrates for solar cell applications.

- Indium is a dopant used for long-wavelength infrared photoconduction silicon detectors in the 3–5 μm atmospheric window.

- Donors, n-type

  - Phosphorus is a n-type dopant. It diffuses fast, so is usually used for bulk doping, or for well formation. Used in solar cells. Can be added by diffusion of phosphine gas. Bulk doping can be achieved by nuclear transmutation, by irradiation of pure silicon with neutrons in a nuclear reactor. Phosphorus also traps gold atoms, which otherwise quickly diffuse through silicon and act as recombination centers.

  - Arsenic is a n-type dopant. Its slower diffusion allows using it for diffused junctions. Used for buried layers. Has similar atomic radius to silicon, high concentrations can be achieved. Its diffusivity is about a tenth of phosphorus or boron, so is used where the dopant should stay in place during subsequent thermal processing. Useful for shallow diffusions where well-controlled abrupt boundary is desired. Preferred dopant in VLSI circuits. Preferred dopant in low resistivity ranges.

  - Antimony is a n-type dopant. It has a small diffusion coefficient. Used for buried layers. Has diffusivity similar to arsenic, is used as its alternative. Its diffusion is virtually purely substitutional, with no interstitials, so it is free of anomalous effects. For this superior property, it is sometimes used in VLSI instead of arsenic. Heavy doping with antimony is important for power devices. Heavily antimony-doped silicon has lower concentration of oxygen impurities; minimal autodoping effects make it suitable for epitaxial substrates.

  - Bismuth is a promising dopant for long-wavelength infrared photoconduction silicon detectors, a viable n-type alternative to the p-type gallium-doped material.

  - Lithium is used for doping silicon for radiation hardened solar cells. The lithium presence anneals defects in the lattice produced by protons and neutrons. Lithium can be introduced to boron-doped p+ silicon, in amounts low enough to maintain the p character of the material, or in large enough amount to counterdope it to low-resistivity n type.

- Other

  - Germanium can be used for band gap engineering. Germanium layer also inhibits diffusion of boron during the annealing steps, allowing ultrashallow p-MOSFET junctions. Germanium bulk doping suppresses large void defects, increases internal gettering, and improves wafer mechanical strength.

  - Silicon, germanium and xenon can be used as ion beams for pre-amorphization of silicon wafer surfaces. Formation of an amorphous layer beneath the surface allows forming ultrashallow junctions for p-MOSFETs.

  - Nitrogen is important for growing defect-free silicon crystal. Improves mechanical strength of the lattice, increases bulk microdefect generation, suppresses vacancy agglomeration.

○ Gold and platinum are used for minority carrier lifetime control. They are used in some infrared detection applications. Gold introduces a donor level 0.35 eV above the valence band and an acceptor level 0.54 eV below the conduction band. Platinum introduces a donor level also at 0.35 eV above the valence band, but its acceptor level is only 0.26 eV below conduction band; as the acceptor level in n-type silicon is shallower, the space charge generation rate is lower and therefore the leakage current is also lower than for gold doping. At high injection levels platinum performs better for lifetime reduction. Reverse recovery of bipolar devices is more dependent on the low-level lifetime, and its reduction is better performed by gold. Gold provides a good tradeoff between forward voltage drop and reverse recovery time for fast switching bipolar devices, where charge stored in base and collector regions must be minimized. Conversely, in many power transistors a long minority carrier lifetime is required to achieve good gain, and the gold/platinum impurities must be kept low.

Other Semiconductors

- Gallium arsenide

    ○ n-type: tellurium, sulphur (substituting As), tin, silicon, germanium (substituting Ga).

    ○ p-type: beryllium, zinc, chromium (substituting Ga), silicon, germanium (substituting As).

- Gallium phosphide

    ○ n-type: tellurium, selenium, sulphur (substituting phosphorus).

    ○ p-type: zinc, magnesium (substituting Ga), tin (substituting P).

- Gallium nitride, Indium gallium nitride, Aluminium gallium nitride

    ○ n-type: silicon (substituting Ga), germanium (substituting Ga, better lattice match), carbon (substituting Ga, naturally embedding into MOVPE-grown layers in low concentration).

    ○ p-type: magnesium (substituting Ga) - challenging due to relatively high ionisation energy above the valence band edge, strong diffusion of interstitial Mg, hydrogen complexes passivating of Mg acceptors and by Mg self-compensation at higher concentrations).

- Cadmium telluride

    ○ n-type: indium, aluminium (substituting Cd), chlorine (substituting Te).

    ○ p-type: phosphorus (substituting Te), lithium, sodium (substituting Cd).

- Cadmium sulfide

    ○ n-type: gallium (substituting Cd), iodine, fluorine (substituting S).

    ○ p-type: lithium, sodium (substituting Cd).

Compensation

In most cases many types of impurities will be present in the resultant doped semiconductor. If an equal number of donors and acceptors are present in the semiconductor, the extra core electrons provided by the former will be used to satisfy the broken bonds due to the latter, so that doping produces no free carriers of either type. This phenomenon is known as compensation, and occurs at the p-n junction in the vast majority of semiconductor devices. Partial compensation, where donors outnumber acceptors or vice versa, allows device makers to repeatedly reverse (invert) the type of a given portion of the material by applying successively higher doses of dopants, so-called counterdoping. Most modern semiconductors are made by successive selective counterdoping steps to create the necessary P and N type areas.

Although compensation can be used to increase or decrease the number of donors or acceptors, the electron and hole mobility is always decreased by compensation because mobility is affected by the sum of the donor and acceptor ions.

Doping in Conductive Polymers

Conductive polymers can be doped by adding chemical reactants to oxidize, or sometimes reduce, the system so that electrons are pushed into the conducting orbitals within the already potentially conducting system. There are two primary methods of doping a conductive polymer, both of which use an oxidation-reduction (i.e., redox) process:

1. Chemical doping involves exposing a polymer such as melanin, typically a thin film, to an oxidant such as iodine or bromine. Alternatively, the polymer can be exposed to a reductant; this method is far less common, and typically involves alkali metals.

2. Electrochemical doping involves suspending a polymer-coated, working electrode in an electrolyte solution in which the polymer is insoluble along with separate counter and reference electrodes. An electric potential difference is created between the electrodes that causes a charge and the appropriate counter ion from the electrolyte to enter the polymer in the form of electron addition (i.e., n-doping) or removal (i.e., p-doping).

N-doping is much less common because the Earth's atmosphere is oxygen-rich, thus creating an oxidizing environment. An electron-rich, n-doped polymer will react immediately with elemental oxygen to de-dope (i.e., reoxidize to the neutral state) the polymer. Thus, chemical n-doping must be performed in an environment of inert gas (e.g., argon). Electrochemical n-doping is far more common in research, because it is easier to exclude oxygen from a solvent in a sealed flask. However, it is unlikely that n-doped conductive polymers are available commercially.

Doping in Organic Molecular Semiconductors

Molecular dopants are preferred in doping molecular semiconductors due to their compatibilities of processing with the host, that is, similar evaporation temperatures or controllable solubility. Additionally, the relatively large sizes of molecular dopants compared with those of metal ion dopants (such as Li^+ and Mo^{6+}) are generally beneficial, yielding excellent spatial confinement for use in multilayer structures, such as OLEDs and Organic solar cells. Typical p-type dopants include F4-

TCNQ and Mo(tfd)$_3$. However, similar to the problem encountered in doping conductive polymers, air-stable n-dopants suitable for materials with low electron affinity (EA) are still elusive. Recently, photoactivation with a combination of cleavable dimeric dopants, such as [RuCp*Mes]$_2$, suggests a new path to realize effective n-doping in low-EA materials.

Magnetic Doping

Research on magnetic doping has shown that considerable alteration of certain properties such as specific heat may be affected by small concentrations of an impurity; for example, dopant impurities in semiconducting ferromagnetic alloys can generate different properties as first predicted by White, Hogan, Suhl and Nakamura. The inclusion of dopant elements to impart dilute magnetism is of growing significance in the field of Magnetic semiconductors. The presence of disperse ferromagnetic species is key to the functionality of emerging Spintronics, a class of systems that utilise electron spin in addition to charge. Using Density functional theory (DFT) the temperature dependent magnetic behaviour of dopants within a given lattice can be modeled to identify candidate semiconductor systems.

Single Dopants in Semiconductors

The sensitive dependence of a semiconductor's properties on dopants has provided an extensive range of tunable phenomena to explore and apply to devices. It is possible to identify the effects of a solitary dopant on commercial device performance as well as on the fundamental properties of a semiconductor material. New applications have become available that require the discrete character of a single dopant, such as single-spin devices in the area of quantum information or single-dopant transistors. Dramatic advances in the past decade towards observing, controllably creating and manipulating single dopants, as well as their application in novel devices have allowed opening the new field of solotronics (solitary dopant optoelectronics).

Neutron Transmutation Doping

Neutron transmutation doping (NTD) is an unusual doping method for special applications. Most commonly, it is used to dope silicon n-type in high-power electronics. It is based on the conversion of the Si-30 isotope into phosphorus atom by neutron absorption as follows:

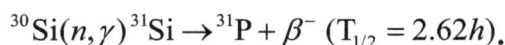

$$^{30}\text{Si}(n,\gamma)\,^{31}\text{Si} \rightarrow\,^{31}\text{P} + \beta^- \ (\text{T}_{1/2} = 2.62h).$$

In practice, the silicon is typically placed near a nuclear reactor to receive the neutrons. As neutrons continue to pass through the silicon, more and more phosphorus atoms are produced by transmutation, and therefore the doping becomes more and more strongly n-type. NTD is a far less common doping method than diffusion or ion implantation, but it has the advantage of creating an extremely uniform dopant distribution.

Modulation Doping

Modulation doping is a synthesis technique in which the dopants are spatially separated from the carriers. In this way, carrier-donor scattering is suppressed, allowing very high mobility to be attained.

HETEROJUNCTION

A heterojunction is the interface that occurs between two layers or regions of dissimilar crystalline semiconductors. These semiconducting materials have unequal band gaps as opposed to a homo-junction. It is often advantageous to engineer the electronic energy bands in many solid-state device applications, including semiconductor lasers, solar cells and transistors ("heterotransistors") to name a few. The combination of multiple heterojunctions together in a device is called a heterostructure, although the two terms are commonly used interchangeably. The requirement that each material be a semiconductor with unequal band gaps is somewhat loose, especially on small length scales, where electronic properties depend on spatial properties. A more modern definition of heterojunction is the interface between any two solid-state materials, including crystalline and amorphous structures of metallic, insulating, fast ion conductor and semiconducting materials.

In 2000, the Nobel Prize in physics was awarded jointly to Herbert Kroemer of the University of California, Santa Barbara, California, USA and Zhores I. Alferov of Ioffe Institute, Saint Petersburg, Russia for "developing semiconductor heterostructures used in high-speed-photography and opto-electronics".

Manufacture and Applications

Heterojunction manufacturing generally requires the use of molecular beam epitaxy (MBE) or chemical vapor deposition (CVD) technologies in order to precisely control the deposition thickness and create a cleanly lattice-matched abrupt interface.

Despite their expense, heterojunctions have found use in a variety of specialized applications where their unique characteristics are critical:

- Lasers: Using heterojunctions in lasers was first proposed in 1963 when Herbert Kroemer, a prominent scientist in this field, suggested that population inversion could be greatly enhanced by heterostructures. By incorporating a smaller direct band gap material like GaAs between two larger band gap layers like AlAs, carriers can be confined so that lasing can occur at room temperature with low threshold currents. It took many years for the material science of heterostructure fabrication to catch up with Kroemer's ideas but now it is the industry standard. It was later discovered that the band gap could be controlled by taking advantage of the quantum size effects in quantum well heterostructures. Furthermore, heterostructures can be used as waveguides to the index step which occurs at the interface, another major advantage to their use in semiconductor lasers. Semiconductor diode lasers used in CD and DVD players and fiber optic transceivers are manufactured using alternating layers of various III-V and II-VI compound semiconductors to form lasing heterostructures.

- Bipolar transistors: When a heterojunction is used as the base-emitter junction of a bipolar junction transistor, extremely high forward gain and low reverse gain result. This translates into very good high frequency operation (values in tens to hundreds of GHz) and low leakage currents. This device is called a heterojunction bipolar transistor (HBT).

- Field effect transistors: Heterojunctions are used in high electron mobility transistors (HEMT) which can operate at significantly higher frequencies (over 500 GHz). The proper doping profile and band alignment gives rise to extremely high electron mobilities by creating a two dimensional electron gas within a dopant free region where very little scattering can occur.

Energy Band Alignment

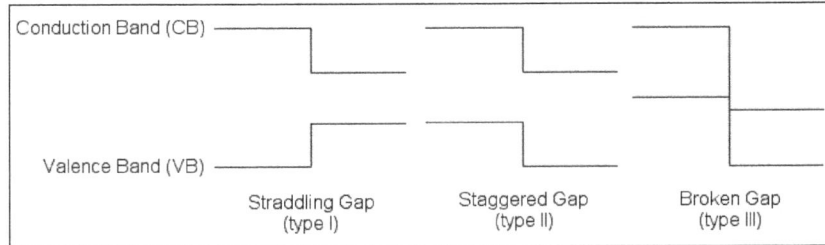

The three types of semiconductor heterojunctions organized by band alignment.

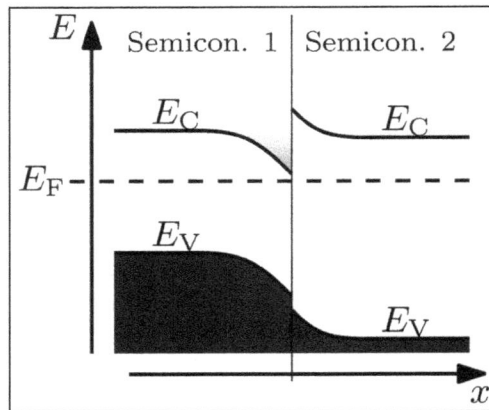

Band diagram for straddling gap, *n-n* semiconductor heterojunction at equilibrium.

The behaviour of a semiconductor junction depends crucially on the alignment of the energy bands at the interface. Semiconductor interfaces can be organized into three types of heterojunctions: straddling gap (type I), staggered gap (type II) or broken gap (type III) as seen in the figure. Away from the junction, the band bending can be computed based on the usual procedure of solving Poisson's equation.

Various models exist to predict the band alignment:

- The simplest (and least accurate) model is Anderson's rule, which predicts the band alignment based on the properties of vacuum-semiconductor interfaces (in particular the vacuum electron affinity). The main limitation is its neglect of chemical bonding.

- A common anion rule was proposed which guesses that since the valence band is related to anionic states, materials with the same anions should have very small valence band offsets. This however did not explain the data but is related to the trend that two materials with different anions tend to have larger valence band offsets than conduction band offsets.

- Tersoff proposed a gap state model based on more familiar metal-semiconductor junctions where the conduction band offset is given by the difference in Schottky barrier height. This model includes a dipole layer at the interface between the two semiconductors which arises from electron tunneling from the conduction band of one material into the gap of the other (analogous to metal-induced gap states). This model agrees well with systems where both materials are closely lattice matched such as GaAs/AlGaAs.

- The *60:40 rule* is a heuristic for the specific case of junctions between the semiconductor GaAs and the alloy semiconductor $Al_xGa_{1-x}As$. As the x in the $Al_xGa_{1-x}As$ side is varied from 0 to 1, the ratio $\Delta E_C / \Delta E_V$ tends to maintain the value 60/40. For comparison, Anderson's rule predicts $\Delta E_C / \Delta E_V = 0.73 / 0.27$ for a GaAs/AlAs junction (x=1).

The typical method for measuring band offsets is by calculating them from measuring exciton energies in the luminescence spectra.

Effective Mass Mismatch

When a heterojunction is formed by two different semiconductors, a quantum well can be fabricated due to difference in band structure. In order to calculate the static energy levels within the achieved quantum well, understanding variation or mismatch of the effective mass across the heterojunction becomes substantial. The quantum well defined in the heterojunction can be treated as a finite well potential with width of l_w. Addition to that, in 1966, Conley et al. and BenDaniel and Duke reported boundary condition for the envelope function in quantum well, known as Ben-Daniel-Duke boundary condition. According to them, the envelope function in fabricated quantum well must satisfy boundary condition which states $\psi(z), \dfrac{1}{m^*}\dfrac{\partial}{\partial z}\psi(z)$ are both continuous in interface regions.

Nanoscale Heterojunctions

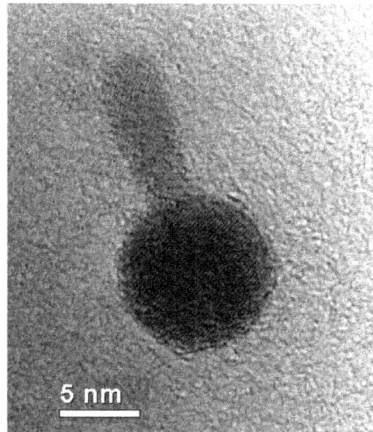

Image of a nanoscale heterojunction between iron oxide (Fe_3O_4 — sphere) and cadmium sulfide (CdS — rod) taken with a TEM. This staggered gap (type II) offset junction was synthesized by Hunter McDaniel and Dr. Moonsub Shim at the University of Illinois in Urbana-Champaign in 2007.

In quantum dots the band energies are dependent on crystal size due to the quantum size effects. This enables band offset engineering in nanoscale heterostructures. It is possible to use the same materials but change the type of junction, say from straddling (type I) to staggered (type II), by changing the size or thickness of the crystals involved. The most common nanoscale heterostructure system is ZnS on CdSe (CdSe@ZnS) which has a straddling gap (type I) offset. In this system the much larger band gap ZnS passivates the surface of the fluorescent CdSe core thereby increasing the quantum efficiency of the luminescence. There is an added bonus of increased thermal stability due to the stronger bonds in the ZnS shell as suggested by its larger band gap. Since CdSe and ZnS both grow in the zincblende crystal phase and are closely lattice matched, core shell growth is preferred. In other systems or under different growth conditions it may be possible to grow anisotropic structures such as the one seen in the image on the right.

It has been shown that the driving force for charge transfer between conduction bands in these structures is the conduction band offset. By decreasing the size of CdSe nanocrystals grown on TiO_2, Robel et al. found that electrons transferred faster from the higher CdSe conduction band into TiO_2. In CdSe the quantum size effect is much more pronounced in the conduction band due to the smaller effective mass than in the valence band, and this is the case with most semiconductors. Consequently, engineering the conduction band offset is typically much easier with nanoscale heterojunctions. For staggered (type II) offset nanoscale heterojunctions, photoinduced charge separation can occur since there the lowest energy state for holes may be on one side of the junction whereas the lowest energy for electrons is on the opposite side. It has been suggested that anisotropic staggered gap (type II) nanoscale heterojunctions may be used for photocatalysis, specifically for water splitting with solar energy.

BAND OFFSET

Band offset describes the relative alignment of the energy bands at a semiconductor heterojunction.

At semiconductor heterojunctions, energy bands of two different materials come together, leading to an interaction. Both band structures are positioned discontinuously from each other, causing them to align close to the interface. This is done to ensure that the Fermi energy level stays continuous throughout the two semiconductors. This alignment is caused by the discontinuous band structures of the semiconductors when compared to each other and the interaction of the two surfaces at the interface.This relative alignment of the energy bands at such semiconductor heterojunctions is called the Band offset.

The band offsets can be determined by both intrinsic properties, that is, determined by properties of the bulk materials, as well as non-intrinsic properties, namely, specific properties of the interface. Depending on the type of the interface, the offsets can be very accurately considered intrinsic, or be able to be modified by manipulating the interfacial structure. Isovalent heterojunctions are generally insensitive to manipulation of the interfacial structure, whilst heterovalent heterojunctions can be influenced in their band offsets by the geometry, the orientation, and the bonds of the interface and the charge transfer between the heterovalent bonds. The band offsets, especially those at heterovalent heterojunctions depend significantly on the distribution of interface charge.

The band offsets are determined by two kinds of factors for the interface, the band discontinuities and the built-in potential. These discontinuities are caused by the difference in band gaps of the semiconductors and are distributed between two band discontinuities, the valence-band discontinuity, and the conduction-band discontinuity. The built-in potential is caused by the bands which bend close at the interface due to a charge imbalance between the two semiconductors, and can be described by Poisson's equation.

Semiconductor Types

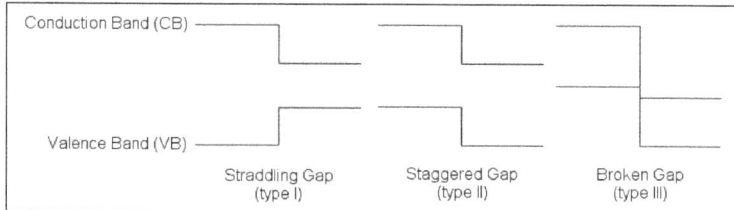

Here is showcased the different types of heterojunctions in semiconductors. In type I, the conduction band of the second semiconductor is lower than that of the first, whilst its valence band is higher than that of the first. As a consequence the band gap of the first semiconductor is larger than the band gap of the second semiconductor. In type II the conduction band and valence band of the second semiconductor are both lower than the bands of the first semiconductor. In this staggered gap, the band gap of the second semiconductor is no longer restricted to being smaller than the first semiconductor, although the band gap of the second semiconductor is still partially contained in the first semiconductor. In type III however, the conduction band of the second semiconductor overlaps with the valence band of the first semiconductor. Due to this overlap, there are no forbidden energies at the interface, and the band gap of the second semiconductor is no longer contained by the band gap of the first.

The behaviour of semiconductor heterojunctions depend on the alignment of the energy bands at the interface and thus on the band offsets. The interfaces of such heterojunctions can be categorized in three types: straddling gap (referred to as type I), staggered gap (type II), and broken gap (type III).

These representations do not take into account the band bending, which is a reasonable assumption if you only look at the interface itself, as band bending exerts its influence on a length scale of generally hundreds of angström. For a more accurate picture of the situation at hand, the inclusion of band bending is important.

In this heterojunction of type I alignment, one can clearly see the built-in potential $\Phi bi = \Phi(A) + \Phi(B)$. The band gap difference $\Delta Eg = Eg(A) - Eg(B)$ is distributed between the two discontinuities, ΔEv, and $\Delta Ec\$$. In alignments, it is generally the case that the conduction band which has the higher energy minimum will bend upward, whilst the valence band which has the lower energy maximum will bend upward. In this type of alignment, this means that both of the bands of semiconductor A will bend upwards, whilst both of the bands of semiconductor B will bend downwards. The band bending, caused by the built-in potential, is determined by the interface position of the Fermi level, and predicting or measuring this level is related to the Schottky barrier height in metal-semiconductor interfaces. Depending on the doping of the bulk material, the band bending can be into the thousands of angstroms, or just fifty, depending on the doping. The discontinuities on the other hand, are primarily due to the electrostatic potential gradients of the abrupt interface, working on a length scale of ideally a single atomic interplanar spacing, and is almost independent of any doping used.

Experimental Methods

Two kinds of experimental techniques are used to describe band offsets. The first is an older technique, the first technique to probe the heterojunction built-in potential and band discontinuities. This methods are generally called transport methods. These methods consist of two classes, either capacitance-voltage (C-V) or current-voltage (I-V) techniques. These older techniques were used to extract the built-in potential by assuming a square-root dependence for the capacitance C on Φ_{bi} - qV, with $_{bi}$ the built-in potential, q the electron charge, and V the applied voltage. If band extrema away from the interface, as well as the distance between the Fermi level, are known parameters, known a priori from bulk doping, it becomes possible to obtain the conduction band offset and the valence band offset. This square root dependence corresponds to an ideally abrupt transition at the interface and it may or may not be a good approximation of the real junction behaviour.

The second kind of technique consists of optical methods. Photon absorption is used effectively as the conduction band and valence band discontinuities define quantum wells for the electrons and the holes. Optical techniques can be used to probe the direct transitions between sub-bands within the quantum wells, and with a few parameters known, such as the geometry of the structure and the effective mass, the transition energy measured experimentally can be used to probe the well depth. Band offset values are usually estimated using the optical response as a function of certain geometrical parameters or the intensity of an applied magnetic field. Light scattering could also be used to determine the size of the well depth.

Alignment

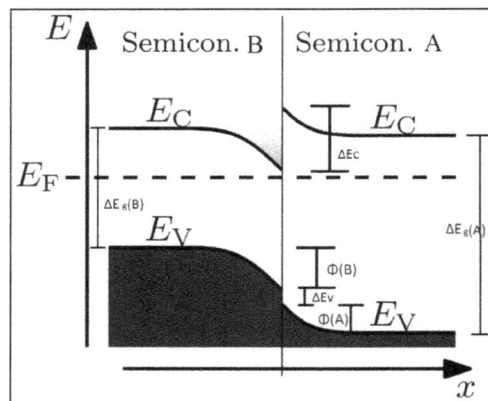

Prediction of the band alignment is at face value dependent on the heterojunction type, as well as whether or not the heterojunction in question is heterovalent or isovalent. However, quantifying this alignment proved a difficult task for a long time. Anderson's rule is used to construct energy band diagrams at heterojunctions between two semiconductors. It states that during the construction of an energy band diagram, the vacuum levels of the semiconductors on either side of the heterojunction should be equal.

Anderson's rule states that when we construct the heterojunction, we need to have both semiconductors on an equal vacuum energy level. This ensures that the energy bands of both the semiconductors are being held to the same reference point, from which ΔE_c and ΔE_v, the conduction band offset and valence band offset can be calculated. By having the same reference point for both semiconductors, ΔE_c becomes equal to the built-in potential, $V_{bi} = \Phi_1 - \Phi_2$, and the behaviour of the bands at the interface can be predicted as can be seen at the picture above.

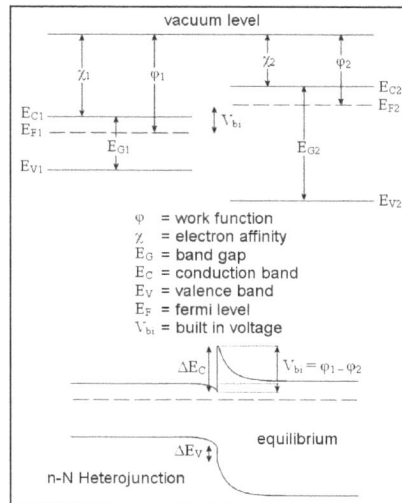

Anderson's rule fails to predict real band offsets. This is primarily due to the fact that Anderson's model implies that the materials are assumed to behave the same as if they were separated by a large vacuum distance, however at these heterojunctions consisting of solids filling the space, there is no vacuum, and the use of the electron affinities at vacuum leads to wrong results. Anderson's rule ignores actual chemical bonding effects that occur on small vacuum separation or non-existent vacuum separation, which leads to wrong predictions about the band offsets.

A better theory for predicting band offsets has been linear-response theory. In this theory, interface dipoles have a significant impact on the lining up of the bands of the semiconductors. These interface dipoles however are not ions, rather they are mathematical constructs based upon the difference of charge density between the bulk and the interface. Linear-response theory is based on first-principles calculations, which are calculations aimed at solving the quantum-mechanical equations, without input from experiment. In this theory, the band offset is the sum of two terms, the first term is intrinsic and depends solely on the bulk properties, the second term, which vanishes for isovalent and abrupt non-polar heterojunctions, depends on the interface geometry, and can easily be calculated once the geometry is known, as well as certain quantities (such as the lattice parameters).

The goal of the model is to attempt to model the difference between the two semiconductors, that is, the difference with respect to an chosen optimal average (whose contribution to the band offset should vanish). An example would be GaAs-AlAs, constructing it from a virtual crystal of $Al_{0.5}Ga_{0.5}As$, then introducing an interface. After this a perturbation is added to turn the crystal into pure GaAs, whilst on the other side, the perturbation transforms the crystal in pure AlAs. These perturbations are sufficiently small so that they can be handled by linear-response theory and the electrostatic potential lineup across the interface can then be obtained up to the first order from the charge density response to those localized perturbations. Linear response theory works well for semiconductors with similar potentials (such as GaAs-AlAs) as well as dissimilar potentials (such as GaAs-Ge), which was doubted at first. However predictions made by linear response theory coincide exactly with those of self-consistent first principle calculations. If interfaces are polar however, or nonabrupt nonpolar oriented, additional effects must be taken into account. These are additional terms which require simple electrostatics, which is within the linear response approach.

FIELD EFFECT

In physics, the field effect refers to the modulation of the electrical conductivity of a material by the application of an external electric field.

In a metal, the electron density that responds to applied fields is so large that an external electric field can penetrate only a very short distance into the material. However, in a semiconductor the lower density of electrons (and possibly holes) that can respond to an applied field is sufficiently small that the field can penetrate quite far into the material. This field penetration alters the conductivity of the semiconductor near its surface, and is called the *field effect*. The field effect underlies the operation of the Schottky diode and of field-effect transistors, notably the MOSFET, the JFET and the MESFET.

Surface Conductance and Band Bending

The change in surface conductance occurs because the applied field alters the energy levels available to electrons to considerable depths from the surface, and that in turn changes the occupancy of the energy levels in the surface region. A typical treatment of such effects is based upon a band-bending diagram showing the positions in energy of the band edges as a function of depth into the material.

An example band-bending diagram is shown in the figure. For convenience, energy is expressed in eV and voltage is expressed in volts, avoiding the need for a factor q for the elementary charge. In the figure, a two-layer structure is shown, consisting of an insulator as left-hand layer and a semiconductor as right-hand layer. An example of such a structure is the *MOS capacitor*, a two-terminal structure made up of a metal *gate* contact, a semiconductor *body* (such as silicon) with a body contact, and an intervening insulating layer (such as silicon dioxide, hence the designation *O*). The left panels show the lowest energy level of the conduction band and the highest energy level of the valence band. These levels are "bent" by the application of a positive voltage V. By convention, the energy of electrons is shown, so a positive voltage penetrating the surface *lowers* the conduction edge. A dashed line depicts the occupancy situation: below this Fermi level the states are more likely to be occupied, the conduction band moves closer to the Fermi level, indicating more electrons are in the conducting band near the insulator.

Bulk Region

The example in the figure shows the Fermi level in the bulk material beyond the range of the applied field as lying close to the valence band edge. This position for the occupancy level is arranged by introducing impurities into the semiconductor. In this case the impurities are so-called *acceptors* which soak up electrons from the valence band becoming negatively charged, immobile ions embedded in the semiconductor material. The removed electrons are drawn from the valence band levels, leaving vacancies or *holes* in the valence band. Charge neutrality prevails in the field-free region because a negative acceptor ion creates a positive deficiency in the host material: a hole is the absence of an electron, it behaves like a positive charge. Where no field is present, neutrality is achieved because the negative acceptor ions exactly balance the positive holes.

Surface Region

Next the band bending is described. A positive charge is placed on the left face of the insulator (for example using a metal "gate" electrode). In the insulator there are no charges so the electric field is constant, leading to a linear change of voltage in this material. As a result, the insulator conduction and valence bands are therefore straight lines in the figure, separated by the large insulator energy gap.

In the semiconductor at the smaller voltage shown in the top panel, the positive charge placed on the left face of the insulator lowers the energy of the valence band edge. Consequently, these states are fully occupied out to a so-called depletion depth where the bulk occupancy reestablishes itself because the field cannot penetrate further. Because the valence band levels near the surface are fully occupied due to the lowering of these levels, only the immobile negative acceptor-ion charges are present near the surface, which becomes an electrically insulating region without holes (the depletion layer). Thus, field penetration is arrested when the exposed negative acceptor ion charge balances the positive charge placed on the insulator surface: the depletion layer adjusts its depth enough to make the net negative acceptor ion charge balance the positive charge on the gate.

Inversion

The conduction band edge also is lowered, increasing electron occupancy of these states, but at low voltages this increase is not significant. At larger applied voltages, however, as in the bottom panel, the conduction band edge is lowered sufficiently to cause significant population of these levels in a narrow surface layer, called an *inversion* layer because the electrons are opposite in polarity to the holes originally populating the semiconductor. This onset of electron charge in the inversion layer becomes very significant at an applied *threshold* voltage, and once the applied voltage exceeds this value charge neutrality is achieved almost entirely by addition of electrons to the inversion layer rather than by an increase in acceptor ion charge by expansion of the depletion layer. Further field penetration into the semiconductor is arrested at this point, as the electron density increases exponentially with band-bending beyond the threshold voltage, effectively *pinning* the depletion layer depth at its value at threshold voltages.

SCHOTTKY BARRIER

A Schottky barrier refers to a metal-semiconductor contact having a large barrier height (i.e. $\phi B > kT$ and low doping concentration that is less than the density of states in the conduction band or valence band. The potential barrier between the metal and the semiconductor can be identified on an energy band diagram. To construct such a diagram we first consider the energy band diagram of the metal and the semiconductor, and align them using the same vacuum level as shown in figure. As the metal and semiconductor are brought together, the Fermi energies of the two materials must be equal at thermal equilibrium figure.

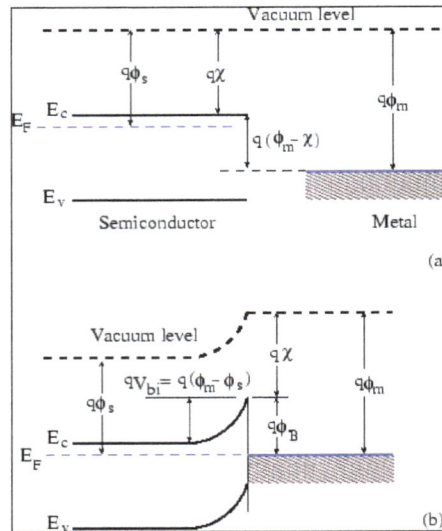

Energy band diagram of a metal adjacent to n-type semiconductor under thermal noneqilibrium condition (a), metal-semiconductor contact in thermal equilibrium (b).

The barrier height ϕB is defined as the potential difference between the Fermi energy of the metal and the band edge where the majority carrier reside. From figure, one finds that for n-type semiconductors the barrier height is obtained from:

$$\phi_{Bn} = \phi_m - X,$$

where ϕ_m is the work function of the metal and X is the electron affinity. The work function of selected metals as measured in vacuum can be found in figure.

For p-type material, the barrier height is given by the difference between the valence band edge and the Fermi energy in the metal:

$$\phi_{B\rho} = \frac{E_g}{q} X - \phi_m,$$

A metal-semiconductor junction will therefore form a barrier for electrons and holes.

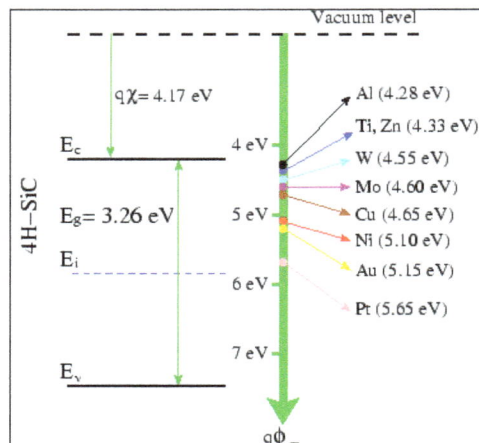

Energy band diagram of the selected metals and 4H-SiC.

If the Fermi energy of the metal is located between the conduction and the valence band edge.

In addition, we define the work function difference as the difference between the work function of the metal and that of the semiconductor. For n-type material it reads:

$$\phi_{wf} = \phi m - X - \frac{E_c - E_{F,n}}{q}$$

similarly, for p-type material:

$$\phi_{wf} = X + \frac{E_c - E_{F,p}}{q} - \phi_m.$$

The work function difference energy becomes:

$$E_w = q.\phi_{wf}$$

The measured barrier height for selected metal/4H-SiC junction is listed in table. These experimental barrier heights depend on the surface polarity of SiC (Si- and C-face), and often differ from the ones calculated above.

Table: Work function of selected metals and their measured and calculated barrier height on n-type 4H-SiC.

	Al	Ti	Zn	W	Mo	Cu	Ni	Au	Pt
ϕm	4.28	4.33	4.33	4.55	4.60	4.65	5.10	5.15	5.65
ϕ_B (Si-face)		1.12					1.69	1.81	
ϕ_B (C-face)		1.25					1.87	2.07	
ϕ_B (calculated)	1.01	1.06	1.06	1.28	1.33	1.38	1.63	1.68	2.08

This is due to the detailed behavior of the metal-semiconductor interface. The ideal metal-semiconductor theory assumes that both materials are pure and that there is no interaction between the two materials nor any interfacial layer. Chemical reactions between the metal and the semiconductor alter the barrier height as do interface states at the surface of the semiconductor and interfacial layers. Furthermore, one finds the barrier heights reported in the literature to vary widely due to different surface cleaning procedures.

The current density is calculated according to the thermionic emission condition neglecting tunneling currents:

$$n \cdot J_n = -q \cdot vn \cdot (n - n_s),$$
$$n \cdot J_p = q \cdot vp \cdot (p - p_s),$$

here, the thermionic recombination velocities v_n and v_p for electrons and holes, respectively are given by:

$$v_v = \sqrt{\frac{k_B \cdot T_L}{2\pi \cdot m_v}} = \frac{4\pi \cdot m_v \cdot (k_B T_L)^2}{N_v \cdot h^3}, \qquad v = n, p$$

and are usually represented by the expression:

$$v_v = A^* \cdot \frac{T_L^2}{q \cdot N_v}, \qquad v = n, p$$

where,

$$A^* = \frac{4\pi \cdot q \cdot m_v \cdot k_B^{\,2}}{h^3}, \qquad v = n, p$$

is known as the effective *Richardson constant*. It is dependent on the effective mass and has a theoretical value of 146 and 72 $Acm^{-2} K^{-2}$ for n-type 4H- and 6H-SiC, respectively.

The carrier concentrations at the surface are given by:

$$n_s = N_c \cdot \exp\left(\frac{-E_c - E_w}{k_B \cdot T_L}\right),$$

$$p_s = N_v \cdot \exp\left(\frac{E_v - E_w}{k_B \cdot T_L}\right).$$

$$J_n - \frac{4\pi q \cdot m_n \cdot (k_B \cdot T_L)^2}{h^3} \exp\left(\frac{q \cdot \phi_m - E_c}{k_B \cdot T_L}\right)\left[\exp\left(\frac{q \cdot \phi_{Bn} - q \cdot \phi_m}{k_B \cdot T_L}\right) - 1\right]$$

$$J_p - \frac{4\pi q \cdot m_p \cdot (k_B \cdot T_L)^2}{h^3} \exp\left(\frac{E_v - q \cdot \phi_m}{k_B \cdot T_L}\right)\left[\exp\left(\frac{q \cdot \phi_m - q \cdot \phi_{Bp}}{k_B \cdot T_L}\right) - 1\right].$$

The Schottky contact boundary conditions for the carrier temperatures T_n and T_p and the lattice temperature T_L are similar to the ones which apply for the Ohmic contact, or respectively.

References

- Neamen, Donald A. (2003). Semiconductor Physics and Devices: Basic Principles (3rd ed.). Mcgraw-Hill Higher Education. ISBN 0-07-232107-5

- Semiconductor, entry: newworldencyclopedia.org, Retrieved 3 March, 2019

- Assadi, M.H.N; Hanaor, D.A.H. (2013). "Theoretical study on copper's energetics and magnetism in

TiO2 polymorphs". Journal of Applied Physics. 113 (23): 233913–233913–5. arXiv:1304.1854. Bibcode:-2013JAP...113w3913A. doi:10.1063/1.4811539

- Node, ayalew: iue.tuwien.ac.at, Retrieved 4 April, 2019

- Ihn, Thomas (2010). "ch. 5.1 Band engineering". Semiconductor Nanostructures Quantum States and Electronic Transport. United States of America: Oxford University Press. P. 66. ISBN 9780199534432

4

Polymers

Polymer is a large molecule made up of various repeated subunits. The important areas of study related to polymers are polymer synthesis, polymer characterization, polymer degradation and copolymers. The topics elaborated in this chapter will help in gaining a better perspective about these aspects of polymers.

A polymer is a large molecule made up of chains or rings of linked repeating subunits, which are called monomers. Polymers usually have high melting and boiling points. Because the molecules consist of many monomers, polymers tend to have high molecular masses.

The word polymer was coined by Swedish chemist Jons Jacob Berzelius in 1833, although with a slightly different meaning from the modern definition. The modern understanding of polymers as macromolecules was proposed by German organic chemist Hermann Staudinger.

Examples of Polymers

Polymers may be divided into two categories. Natural polymers (also called biopolymers) include silk, rubber, cellulose, wool, amber, keratin, collagen, starch, DNA, and shellac. Biopolymers serve key functions in organisms, acting as structural proteins, functional proteins, nucleic acids, structural polysaccharides, and energy storage molecules.

Synthetic polymers are prepared by a chemical reaction, often in a lab. Examples of synthetic polymers include PVC (polyvinyl chloride), polystyrene, synthetic rubber, silicone, polyethylene, neoprene, and nylon. Synthetic polymers are used to make plastics, adhesives, paints, mechanical parts, and many common objects.

Synthetic polymers may be grouped into two categories. Thermoset plastics are made from a liquid or soft solid substance that can be irreversibly changed into an insoluble polymer by curing using heat or radiation. Thermoset plastics tend to be rigid and have high molecular weights. The plastic stays out of shape when deformed and typically decompose before they melt. Examples of thermoset plastics include epoxy, polyester, acrylic resins, polyurethanes, and vinyl esters. Bakelite, Kevlar, and vulcanized rubber are also thermoset plastics.

Thermoplastic polymers or thermosoftening plastics are the other type of synthetic polymers. While thermoset plastics are rigid, thermoplastic polymers are solid when cool, but are pliable and can be molded above a certain temperature. While thermoset plastics form irreversible chemical bonds when cured, the bonding in thermoplastics weakens with temperature. Unlike thermosets, which decompose rather than melt, thermoplastics melt into a liquid upon heating.

Examples of thermoplastics include acrylic, nylon, Teflon, polypropylene, polycarbonate, ABS, and polyethylene.

The Structure of Polymers

Many common classes of polymers are composed of hydrocarbons, compounds of carbon and hydrogen. These polymers are specifically made of carbon atoms bonded together, one to the next, into long chains that are called the backbone of the polymer. Because of the nature of carbon, one or more other atoms can be attached to each carbon atom in the backbone. There are polymers that contain only carbon and hydrogen atoms. Polyethylene, polypropylene, polybutylene, polystyrene and polymethylpentene are examples of these. Polyvinyl chloride (PVC) has chlorine attached to the all-carbon backbone. Teflon has fluorine attached to the all-carbon backbone.

Other common manufactured polymers have backbones that include elements other than carbon. Nylons contain nitrogen atoms in the repeat unit backbone. Polyesters and polycarbonates contain oxygen in the backbone. There are also some polymers that, instead of having a carbon backbone, have a silicon or phosphorous backbone. These are considered inorganic polymers. One of the more famous silicon-based polymers is Silly Putty®.

Molecular Arrangement of Polymers

Think of how spaghetti noodles look on a plate. These are similar to how linear polymers can be arranged if they lack specific order, or are amorphous. Controlling the polymerization process and quenching molten polymers can result in amorphous organization. An amorphous arrangement of molecules has no long-range order or form in which the polymer chains arrange themselves. Amorphous polymers are generally transparent. This is an important characteristic for many applications such as food wrap, plastic windows, headlight lenses and contact lenses.

Obviously not all polymers are transparent. The polymer chains in objects that are translucent and opaque may be in a crystalline arrangement. By definition, a crystalline arrangement has atoms, ions, or in this case, molecules arranged in distinct patterns. You generally think of crystalline structures in table salt and gemstones, but they can occur in plastics. Just as quenching can produce amorphous arrangements, processing can control the degree of crystallinity for those polymers that are able to crystallize. Some polymers are designed to never be able to crystallize. Others are designed to be able to be crystallized. The higher the degree of crystallinity, generally, the less light can pass through the polymer. Therefore, the degree of translucence or opaqueness of the polymer can be directly affected by its crystallinity. Crystallinity creates benefits in strength, stiffness, chemical resistance, and stability.

Scientists and engineers are always producing more useful materials by manipulating the molecular structure that affects the final polymer produced. Manufacturers and processors introduce various fillers, reinforcements and additives into the base polymers, expanding product possibilities.

Characteristics of Polymers

The majority of manufactured polymers are thermoplastic, meaning that once the polymer is formed it can be heated and reformed over and over again. This property allows for easy processing and facilitates recycling. The other group, the thermosets, cannot be remelted. Once these polymers are formed, reheating will cause the material to ultimately degrade, but not melt.

Every polymer has very distinct characteristics, but most polymers have the following general attributes:

1. Polymers can be very resistant to chemicals. Consider all the cleaning fluids in your home that are packaged in plastic. Reading the warning labels that describe what happens when the chemical comes in contact with skin or eyes or is ingested will emphasize the need for chemical resistance in the plastic packaging. While solvents easily dissolve some plastics, other plastics provide safe, non-breakable packages for aggressive solvents.

2. Polymers can be both thermal and electrical insulators. A walk through your house will reinforce this concept, as you consider all the appliances, cords, electrical outlets and wiring that are made or covered with polymeric materials. Thermal resistance is evident in the kitchen with pot and pan handles made of polymers, the coffee pot handles, the foam core of refrigerators and freezers, insulated cups, coolers, and microwave cookware. The thermal underwear that many skiers wear is made of polypropylene and the fiberfill in winter jackets is acrylic and polyester.

3. Generally, polymers are very light in weight with significant degrees of strength. Consider the range of applications, from toys to the frame structure of space stations, or from delicate nylon fiber in pantyhose to Kevlar, which is used in bulletproof vests. Some polymers float in water while others sink. But, compared to the density of stone, concrete, steel, copper, or aluminum, all plastics are lightweight materials.

4. Polymers can be processed in various ways. Extrusion produces thin fibers or heavy pipes or films or food bottles. Injection molding can produce very intricate parts or large car body panels. Plastics can be molded into drums or be mixed with solvents to become adhesives or paints. Elastomers and some plastics stretch and are very flexible. Some plastics are stretched in processing to hold their shape, such as soft drink bottles. Other polymers can be foamed like polystyrene (Styrofoam™), polyurethane and polyethylene.

5. Polymers are materials with a seemingly limitless range of characteristics and colors. Polymers have many inherent properties that can be further enhanced by a wide range of additives to broaden their uses and applications. Polymers can be made to mimic cotton, silk, and wool fibers; porcelain and marble; and aluminum and zinc. Polymers can also make possible products that do not readily come from the natural world, such as clear sheets and flexible films.

6. Polymers are usually made of petroleum, but not always. Many polymers are made of repeat units derived from natural gas or coal or crude oil. But building block repeat units can sometimes be made from renewable materials such as polylactic acid from corn or cellulosics from cotton linters. Some plastics have always been made from renewable materials such as cellulose acetate used for screwdriver handles and gift ribbon. When the building blocks can be made more economically from renewable materials than from fossil fuels, either old plastics find new raw materials or new plastics are introduced.

7. Polymers can be used to make items that have no alternatives from other materials. Polymers can be made into clear, waterproof films. PVC is used to make medical tubing and blood bags that extend the shelf life of blood and blood products. PVC safely delivers flammable oxygen in non-burning flexible tubing. And anti-thrombogenic material, such as heparin, can be incorporated into flexible PVC catheters for open heart surgery, dialysis, and blood collection. Many medical devices rely on polymers to permit effective functioning.

Solid Waste Management

In addressing all the superior attributes of polymers, it is equally important to discuss some of the challenges associated with the materials. Most plastics deteriorate in full sunlight, but never decompose completely when buried in landfills. However, other materials such as glass, paper, or aluminum do not readily decompose in landfills either. Some bioplastics do decompose to carbon dioxide and water, however, in specially designed food waste commercial composting facilities only. They do not biodegrade under other circumstances.

For 2005 the EPA characterization of municipal solid waste before recycling for the United States showed plastics made up 11.8 percent of our trash by weight compared to paper that constituted 34.2 percent. Glass and metals made up 12.8 percent by weight. And yard trimmings constituted 13.1 percent of municipal solid waste by weight. Food waste made up 11.9 percent of municipal solid waste. The characteristics that make polymers so attractive and useful, lightweight and almost limitless physical forms of many polymers designed to deliver specific appearance and functionality, make post-consumer recycling challenging. When enough used plastic items can be gathered together, companies develop technology to recycle those used plastics. The recycling rate for all plastics is not as high as any would want. But, the recycling rate for the 1,170,000,000 pounds of polyester bottles, 23.1%, recycled in 2005 and the 953,000,000 pounds of high density polyethylene bottles, 28.8%, recycled in 2005 show that when critical mass of defined material is available, recycling can be a commercial success.

Applications for recycled plastics are growing every day. Recycled plastics can be blended with virgin plastic (plastic that has not been processed before) without sacrificing properties in many applications. Recycled plastics are used to make polymeric timbers for use in picnic tables, fences and outdoor playgrounds, thus providing low maintenance, no splinters products and saving natural lumber. Plastic from soft drink and water bottles can be spun into fiber for the production of carpet or made into new food bottles. Closed loop recycling does occur, but sometimes the most valuable use for a recycled plastic is into an application different than the original use.

An option for plastics that are not recycled, especially those that are soiled, such as used food wrap or diapers, can be a waste-to-energy system (WTE). In 2005, 13.6% of US municipal solid waste was processed in WTE systems. When localities decide to use waste-to-energy systems to manage solid waste, plastics can be a useful component.

The controlled combustion of polymers produces heat energy. The heat energy produced by the burning plastic municipal waste not only can be converted to electrical energy but also helps burn the wet trash that is present. Paper also produces heat when burned, but not as much as do plastics. On the other hand, glass, aluminum and other metals do not release any energy when burned.

To better understand the incineration process, consider the smoke coming off a burning item. If one were to ignite the smoke with a lit propane torch, one would observe that the smoke disappears. This exercise illustrates that the by-products of incomplete burning are still flammable. Proper incineration burns the material and the by-products of the initial burning and also takes care of air and solid emissions to insure public safety.

Some plastics can be composted either because of special additives or because of the construction of the polymers. Compostable plastics frequently require more intense conditions to decompose than are available in backyard compost piles. Commercial composters are suggested for compostable plastics. In 2005, composting processed 8.4% of US municipal solid waste.

Plastics can also be safely land filled, although the valuable energy resource of the plastics would then be lost for recycling or energy capture. In 2005, 54.3% of US municipal solid waste was land filled. Plastics are used to line landfills so that leachate is captured and groundwater is not polluted. Non-degrading plastics help stabilize the ground so that after the landfill is closed, the land can be stable enough for useful futures.

Polymers affect every day of our life. These materials have so many varied characteristics and applications that their usefulness can only be measured by our imagination. Polymers are the materials of past, present and future generations.

POLYMER SYNTHESIS

Polymers are long chain giant organic molecules are assembled from many smaller molecules called monomers. Polymers consist of many repeating monomer units in long chains. A polymer is analogous to a necklace made from many small beads (monomers). Many monomers are alkenes or other molecules with double bonds which react by addition to their unsaturated double bonds.

The electrons in the double bond are used to bond two monomer molecules together. This is represented by the red arrows moving from one molecule to the space between two molecules where a new bond is to form. The formation of polyethylene from ethylene (ethene) may be illustrated in the graphic on the left as follows. In the complete polymer, all of the double bonds have been turned into single bonds. No atoms have been lost and you can see that the monomers have just been joined in the process of addition. A simple representation is -[A-A-A-A-A]-. Polyethylene is used in plastic bags, bottles, toys, and electrical insulation.

- LDPE - Low Density Polyethylene: The first commercial polyethylene process used peroxide catalysts at a temperature of 500 C and 1000 atmospheres of pressure. This yields a transparent polymer with highly branched chains which do not pack together well and is low in density. LDPE makes a flexible plastic. Today most LDPE is used for blow-molding of films for packaging and trash bags and flexible snap-on lids. LDPE is recyclable plastic #4.

- HDPE - High Density Polyethylene: An alternate method is to use Ziegler-Natta aluminum titanium catalysts to make HDPE which has very little branching, allows the strands to pack closely, and thus is high density. It is three times stronger than LDPE and more opaque. About 45% of the HDPE is blow molded into milk and disposable consumer bottles. HDPE is also used for crinkly plastic bags to pack groceries at grocery stores. HDPE is recyclable plastic #2.

Other Addition Polymers

- PVC (polyvinyl chloride), which is found in plastic wrap, simulated leather, water pipes, and garden hoses, is formed from vinyl chloride ($H_2C=CHCl$). The reaction is shown in the graphic on the left. Notice how every other carbon must have a chlorine attached.

- Polypropylene: The reaction to make polypropylene ($H_2C=CHCH_3$) is illustrated in the middle reaction of the graphic. Notice that the polymer bonds are always through the carbons of the double bond. Carbon #3 already has saturated bonds and cannot participate in any new bonds. A methyl group is on every other carbon.

- Polystyrene: The reaction is the same for polystrene where every other carbon has a benzene ring attached. Polystyrene (PS) is recyclable plastic. In the following illustrated example, many styrene monomers are polymerized into a long chain polystyrene molecule. The squiggly lines indicate that the polystyrene molecule extends further at both the left and right ends.

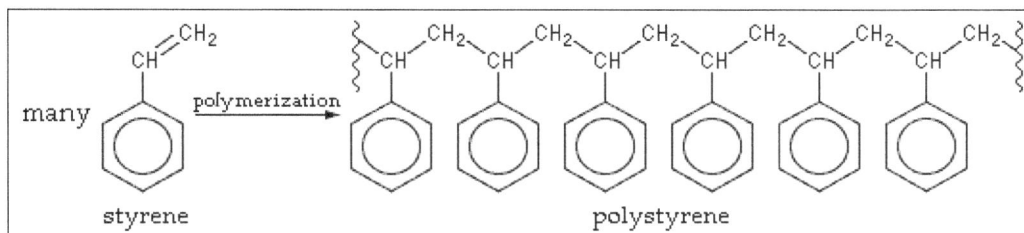

- Blowing fine gas bubbles into liquid polystyrene and letting it solidify produces *expanded polystyrene*, called *Styrofoam* by the Dow Chemical Company.

- Polystyrene with DVB: Cross-linking between polymer chains can be introduced into polystyrene by copolymerizing with *p*-divinylbenzene (DVB). DVB has vinyl groups (-CH=CH$_2$) at each end of its molecule, each of which can be polymerized into a polymer chain like any other vinyl group on a styrene monomer.

Table: Links to various polymers with Chime molecule - Macrogalleria at U. Southern Mississippi.

Monomer	Polymer Name	Trade Name	Uses
$F_2C=CF_2$	polytetrafluoroethylene	Teflon	Non-stick coating for cooking utensils, chemically-resistant specialty plastic parts, Gore-Tex
$H_2C=CCl_2$	polyvinylidene dichloride	Saran	Clinging food wrap

Monomer	Polymer Name	Trade Name	Uses
$H_2C=CH(CN)$	polyacrylonitrile	Orlon, Acrilan, Creslan	Fibers for textiles, carpets, upholstery
$H_2C=CH(OCOCH_3)$	polyvinyl acetate	–	Elmer's glue - Silly Putty Demo
$H_2C=CH(OH)$	polyvinyl alcohol	–	Ghostbusters Demo
$H_2C=C(CH_3)COOCH_3$	polymethyl methacrylate	Plexiglass, Lucite	Stiff, clear, plastic sheets, blocks, tubing, and other shapes

Addition Polymers from Conjugated Dienes

Polymers from conjugated dienes usually give elastomer polymers having rubber-like properties.

Table: Addition homopolymers from conjugated dienes.

Monomer	Polymer Name	Trade Name	Uses
$H_2C=CH-C(CH_3)=CH_2$	polyisoprene	natural or some synthetic rubber	applications similar to natural rubber
$H_2C=CH-CH=CH_2$	polybutadiene	polybutadiene synthetic rubber	select synthetic rubber applications
$H_2C=CH-CCl=CH_2$	polychloroprene	Neoprene	chemically-resistant rubber

All the monomers from which addition polymers are made are alkenes or functionally substituted alkenes. The most common and thermodynamically favored chemical transformations of alkenes are addition reactions. Many of these addition reactions are known to proceed in a stepwise fashion by way of reactive intermediates, and this is the mechanism followed by most polymerizations. A general diagram illustrating this assembly of linear macromolecules, which supports the name chain growth polymers, is presented here. Since a pi-bond in the monomer is converted to a sigma-bond in the polymer, the polymerization reaction is usually exothermic by 8 to 20 kcal/mol. Indeed, cases of explosively uncontrolled polymerizations have been reported.

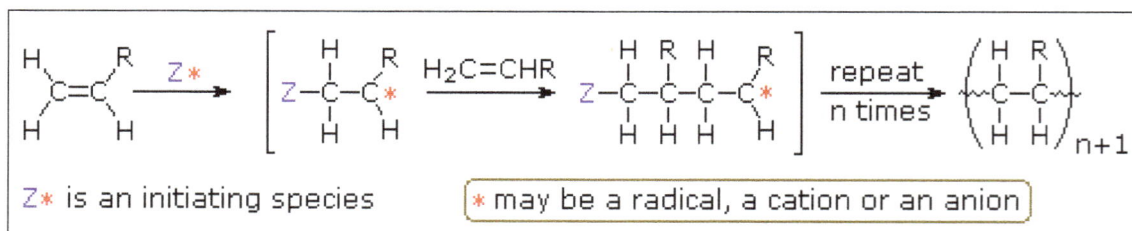

$Z*$ is an initiating species $*$ may be a radical, a cation or an anion

It is useful to distinguish four polymerization procedures fitting this general description.

- Radical Polymerization The initiator is a radical, and the propagating site of reactivity (*) is a carbon radical.

- Cationic Polymerization The initiator is an acid, and the propagating site of reactivity (*) is a carbocation.

- Anionic Polymerization The initiator is a nucleophile, and the propagating site of reactivity (*) is a carbanion.

- Coordination Catalytic Polymerization The initiator is a transition metal complex, and the propagating site of reactivity (*) is a terminal catalytic complex.

Radical Chain-growth Polymerization

Virtually all of the monomers are subject to radical polymerization. Since this can be initiated by traces of oxygen or other minor impurities, pure samples of these compounds are often "stabilized" by small amounts of radical inhibitors to avoid unwanted reaction. When radical polymerization is desired, it must be started by using a radical initiator, such as a peroxide or certain azo compounds. The formulas of some common initiators, and equations showing the formation of radical species from these initiators are presented below.

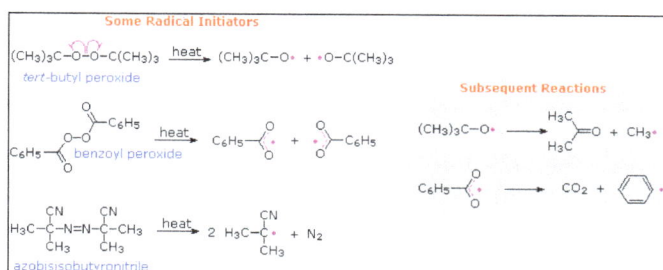

By using small amounts of initiators, a wide variety of monomers can be polymerized. One example of this radical polymerization is the conversion of styrene to polystyrene, shown in the following diagram. The first two equations illustrate the initiation process, and the last two equations are examples of chain propagation. Each monomer unit adds to the growing chain in a manner that generates the most stable radical. Since carbon radicals are stabilized by substituents of many kinds, the preference for head-to-tail regioselectivity in most addition polymerizations is understandable. Because radicals are tolerant of many functional groups and solvents (including water), radical polymerizations are widely used in the chemical industry.

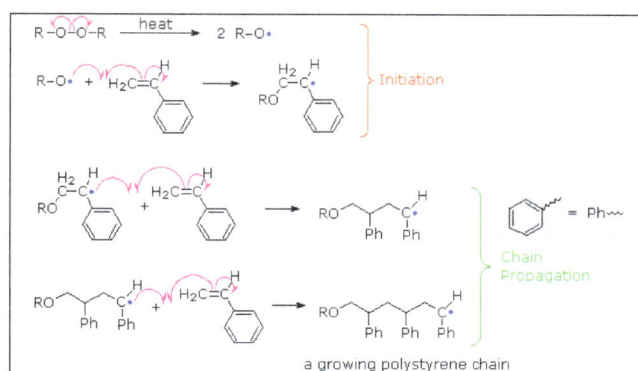

In principle, once started a radical polymerization might be expected to continue unchecked, producing a few extremely long chain polymers. In practice, larger numbers of moderately sized chains are formed, indicating that chain-terminating reactions must be taking place. The most common termination processes are Radical Combination and Disproportionation. These reactions are illustrated by the following equations. The growing polymer chains are colored blue and red, and the hydrogen atom transferred in disproportionation is colored green. Note that in both types of termination two reactive radical sites are removed by simultaneous conversion to stable product(s). Since the concentration of radical species in a polymerization reaction is small relative to other reactants (e.g. monomers, solvents and terminated chains), the rate at which these radical-radical termination reactions occurs is very small, and most growing chains achieve moderate length before termination.

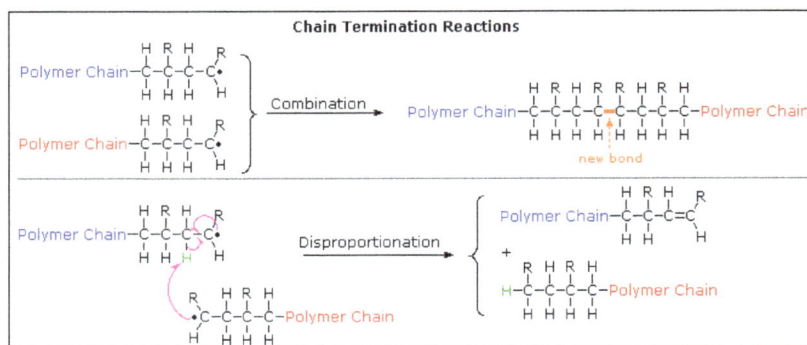

The relative importance of these terminations varies with the nature of the monomer undergoing polymerization. For acrylonitrile and styrene combination is the major process. However, methyl methacrylate and vinyl acetate are terminated chiefly by disproportionation.

Another reaction that diverts radical chain-growth polymerizations from producing linear macromolecules is called chain transfer. As the name implies, this reaction moves a carbon radical from one location to another by an intermolecular or intramolecular hydrogen atom transfer (colored green). These possibilities are demonstrated by the following equations.

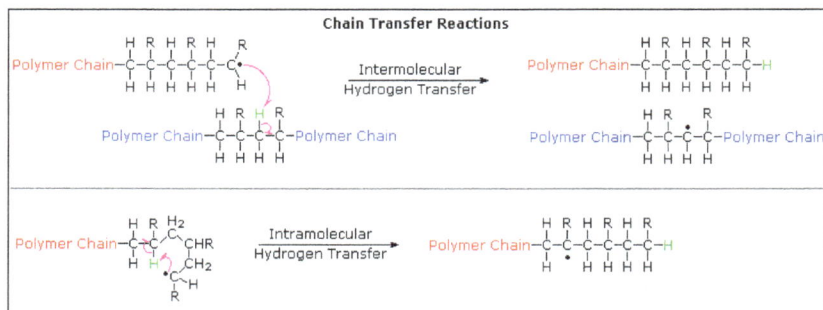

Chain transfer reactions are especially prevalent in the high pressure radical polymerization of ethylene, which is the method used to make LDPE (low density polyethylene). The 1°-radical at the end of a growing chain is converted to a more stable 2°-radical by hydrogen atom transfer. Further polymerization at the new radical site generates a side chain radical, and this may in turn lead to creation of other side chains by chain transfer reactions. As a result, the morphology of LDPE is an amorphous network of highly branched macromolecules.

Chain Topology

Polymers may also be classified as straight-chained or branched, leading to forms such as these:

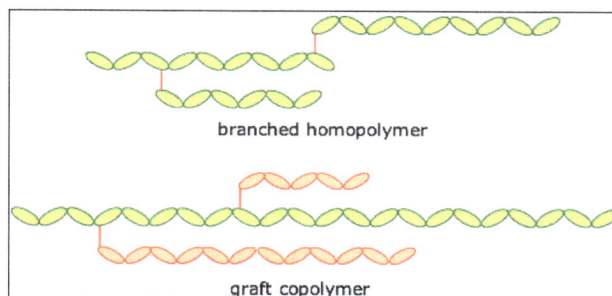

The monomers can be joined end-to-end, and they can also be cross-linked to provide a harder material:

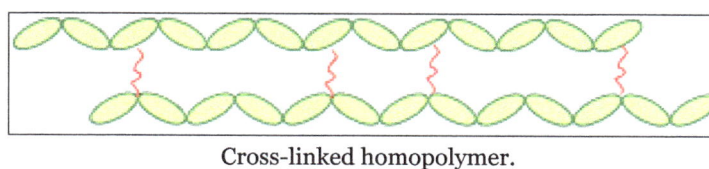

Cross-linked homopolymer.

If the cross-links are fairly long and flexible, adjacent chains can move with respect to each other, producing an *elastic* polymer.

Cationic Chain-growth Polymerization

Polymerization of isobutylene (2-methylpropene) by traces of strong acids is an example of cationic polymerization. The polyisobutylene product is a soft rubbery solid, Tg = _70º C, which is used for inner tubes. This process is similar to radical polymerization, as demonstrated by the following equations. Chain growth ceases when the terminal carbocation combines with a nucleophile or loses a proton, giving a terminal alkene (as shown here).

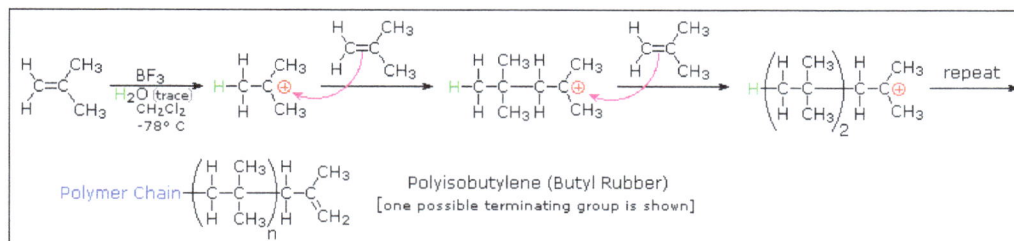

Monomers bearing cation stabilizing groups, such as alkyl, phenyl or vinyl can be polymerized by cationic processes. These are normally initiated at low temperature in methylene chloride solution. Strong acids, such as $HClO_4$, or Lewis acids containing traces of water serve as initiating reagents. At low temperatures, chain transfer reactions are rare in such polymerizations, so the resulting polymers are cleanly linear (unbranched).

Anionic Chain-Growth Polymerization

Treatment of a cold THF solution of styrene with 0.001 equivalents of n-butyllithium causes an immediate polymerization. This is an example of anionic polymerization, the course of which is described by the following equations. Chain growth may be terminated by water or carbon dioxide, and chain transfer seldom occurs. Only monomers having anion stabilizing substituents, such as phenyl, cyano or carbonyl are good substrates for this polymerization technique. Many of the resulting polymers are largely isotactic in configuration, and have high degrees of crystallinity.

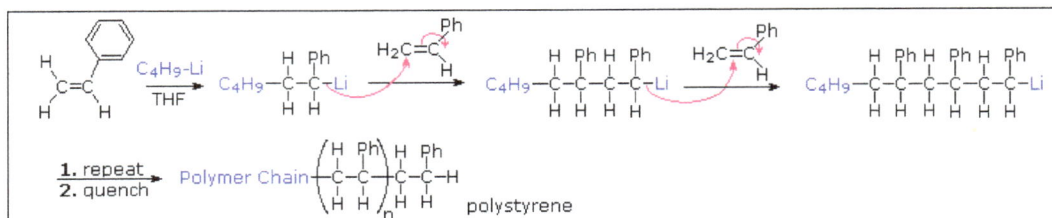

Species that have been used to initiate anionic polymerization include alkali metals, alkali amides, alkyl lithiums and various electron sources. A practical application of anionic polymerization occurs in the use of superglue. This material is methyl 2-cyanoacrylate, $CH_2=C(CN)CO_2CH_3$. When exposed to water, amines or other nucleophiles, a rapid polymerization of this monomer takes place.

Ring Opening Polymerization

In this kind of polymerization, molecular rings are opened in the formation of a polymer. Here epsilon-caprolactam, a 6-carbon cyclic monomer, undergoes ring opening to form a Nylon 6 homopolymer, which is somewhat similar to but not the same as Nylon 6,6 alternating copolymer.

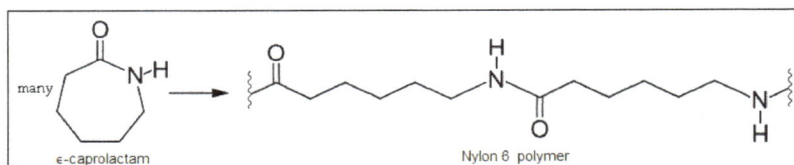

Addition Copolymerization

Most direct copolymerizations of equimolar mixtures of different monomers give statistical copolymers, or if one monomer is much more reactive a nearly homopolymer of that monomer. The copolymerization of styrene with methyl methacrylate, for example, proceeds differently depending on the mechanism. Radical polymerization gives a statistical copolymer. However, the product of cationic polymerization is largely polystyrene, and anionic polymerization favors formation of poly(methyl methacrylate). In cases where the relative reactivities are different, the copolymer composition can sometimes be controlled by continuous introduction of a biased mixture of monomers into the reaction.

Formation of alternating copolymers is favored when the monomers have different polar substituents (e.g. one electron withdrawing and the other electron donating), and both have similar reactivities toward radicals. For example, styrene and acrylonitrile copolymerize in a largely alternating fashion.

Some Useful Copolymers			
Monomer A	Monomer B	Copolymer	Uses
$H_2C=CHCl$	$H_2C=CCl_2$	Saran	films and fibers

H$_2$C=CHC$_6$H$_5$	H$_2$C=C-CH=CH$_2$	SBR styrene butadiene rubber	tires
H$_2$C=CHCN	H$_2$C=C-CH=CH$_2$	Nitrile Rubber	adhesives hoses
H$_2$C=C(CH$_3$)$_2$	H$_2$C=C-CH=CH$_2$	Butyl Rubber	inner tubes
F$_2$C=CF(CF$_3$)	H$_2$C=CHF	Viton	gaskets

A terpolymer of acrylonitrile, butadiene and styrene, called ABS rubber, is used for high-impact containers, pipes and gaskets.

Block Copolymerization

Several different techniques for preparing block copolymers have been developed, many of which use condensation reactions. In the anionic polymerization of styrene , a reactive site remains at the end of the chain until it is quenched. The unquenched polymer has been termed a living polymer, and if additional styrene or a different suitable monomer is added a block polymer will form. This is illustrated for methyl methacrylate in the following diagram.

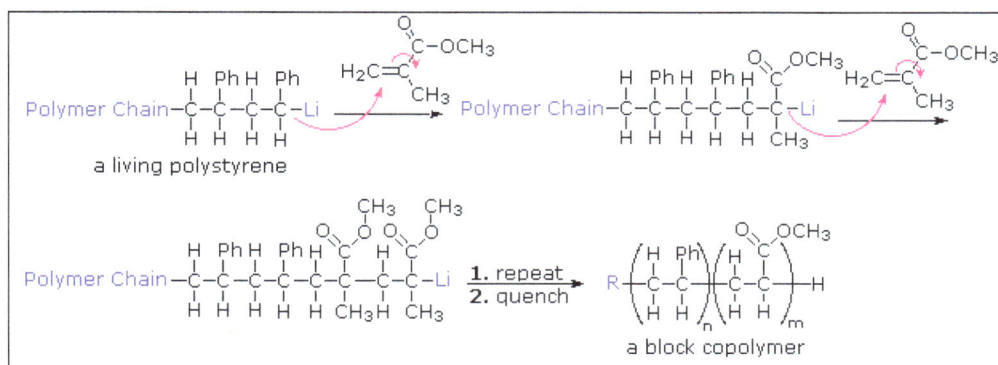

POLYMER CHARACTERIZATION

Polymer characterization is the analytical branch of polymer science. The discipline is concerned with the characterization of polymeric materials on a variety of levels. The characterization typically has as a goal to improve the performance of the material. As such, many characterization techniques should ideally be linked to the desirable properties of the material such as strength, impermeability, thermal stability, and optical properties.

Characterization techniques are typically used to determine molecular mass, molecular structure, morphology, thermal properties, and mechanical properties.

Molecular Mass

The molecular mass of a polymer differs from typical molecules, in that polymerization reactions produce a distribution of molecular weights and shapes. The distribution of molecular masses can be summarized by the number average molecular weight, weight average molecular weight, and polydispersity. Some of the most common methods for determining these parameters are colligative property measurements, static light scattering techniques, viscometry, and size exclusion chromatography.

Gel permeation chromatography, a type of size exclusion chromatography, is an especially useful technique used to directly determine the molecular weight distribution parameters based on the polymer's hydrodynamic volume. Gel permeation chromatography is often used in combination with multi-angle light scattering (MALS), Low-angle laser light scattering (LALLS) and/or viscometry for an absolute determination (i.e., independent of the chromatographic separation details) of the molecular weight distribution as well as the branching ratio and degree of long chain branching of a polymer, provided a suitable solvent can be found.

Molar mass determination of copolymers is a much more complicated procedure. The complications arise from the effect of solvent on the homopolymers and how this can affect the copolymer morphology. Analysis of copolymers typically requires multiple characterization methods. For instance, copolymers with short chain branching such as linear low-density polyethylene (a copolymer of ethylene and a higher alkene such as hexene or octene) require the use of Analytical Temperature Rising Elution Fractionation (ATREF) techniques. These techniques can reveal how the short chain branches are distributed over the various molecular weights. A more efficient analysis of copolymer molecular mass and composition is possible using GPC combined with a triple-detection system comprising multi-angle light scattering, UV absorption and differential refractometry, if the copolymer is composed of two base polymers that provide different responses to UV and/or refractive index.

Molecular Structure

Many of the analytical techniques used to determine the molecular structure of unknown organic compounds are also used in polymer characterization. Spectroscopic techniques such as ultraviolet-visible spectroscopy, infrared spectroscopy, Raman spectroscopy, nuclear magnetic resonance spectroscopy, electron spin resonance spectroscopy, X-ray diffraction, and mass spectrometry are used to identify common functional groups.

Morphology

Polymer morphology is a microscale property that is largely dictated by the amorphous or crystalline portions of the polymer chains and their influence on each other. Microscopy techniques are especially useful in determining these microscale properties, as the domains created by the polymer morphology are large enough to be viewed using modern microscopy instruments. Some of the most common microscopy techniques used are X-ray diffraction, Transmission Electron Microscopy, Scanning Transmission Electron Microscopy, Scanning Electron Microscopy, and Atomic Force Microscopy.

Polymer morphology on a mesoscale (nanometers to micrometers) is particularly important for the mechanical properties of many materials. Transmission Electron Microscopy in combination with staining techniques, but also Scanning Electron Microscopy, Scanning probe microscopy are important tools to optimize the morphology of materials like polybutadiene-polystyrene polymers and many polymer blends.

X-ray diffraction is generally not as powerful for this class of materials as they are either amorphous or poorly crystallized. The Small-angle scattering like Small-angle X-ray scattering (SAXS) can be used to measure the long periods of semicrystalline polymers.

Thermal Properties

A true workhorse for polymer characterization is thermal analysis, particularly Differential scanning calorimetry. Changes in the compositional and structural parameters of the material usually affect its melting transitions or glass transitions and these in turn can be linked to many performance parameters. For semicrystalline polymers it is an important method to measure crystallinity. Thermogravimetric analysis can also give an indication of polymer thermal stability and the effects of additives such as flame retardants. Other thermal analysis techniques are typically combinations of the basic techniques and include differential thermal analysis, thermomechanical analysis, dynamic mechanical thermal analysis, and dielectric thermal analysis.

Dynamic mechanical spectroscopy and dielectric spectroscopy are essentially extensions of thermal analysis that can reveal more subtle transitions with temperature as they affect the complex modulus or the dielectric function of the material.

Mechanical Properties

The characterization of mechanical properties in polymers typically refers to a measure of the strength, elasticity, viscoelasticity, and anisotropy of a polymeric material. The mechanical properties of a polymer are strongly dependent upon the Van der Waals interactions of the polymer chains, and the ability of the chains to elongate and align in the direction of the applied force. Other phenomena, such as the propensity of polymers to form crazes can impact the mechanical properties. Typically, polymeric materials are characterized as elastomers, plastics, or rigid polymers depending their mechanical properties.

The tensile strength, yield strength, and Young's modulus are measures of strength and elasticity, and are of particular interest for describing the stress-strain properties of polymeric materials. These properties can be measure through tensile testing. For crystalline or semicrystalline polymers, anisotropy plays a large role in the mechanical properties of the polymer. The crystallinity of the polymer can be measured through differential scanning calorimetry. For amorphous and semicrystalline polymers, as stress is applied, the polymer chains are able to disentangle and align. If the stress is applied in the direction of chain alignment, the polymer chains will exhibit a higher yield stress and strength, as the covalent bonds connecting the backbone of the polymer absorb the stress. However, if the stress is applied normal to the direction of chain alignment, the Van der Waals interactions between chains will primarily be responsible for the mechanical properties and thus, the yield stress will decrease. This would be observable in a stress strain graph found through tensile testing. Sample preparation, including chain orientation within the sample, for tensile tests therefore can play a large role in the observed mechanical properties.

The fracture properties of crystalline and semicrystalline polymers can be evaluated with Charpy impact testing. Charpy tests, which can also be used with alloy systems, are performed by creating a notch in the sample, and then using a pendulum to fracture the sample at the notch. The pendulum's motion can be used to extrapolate the energy absorbed by the sample to fracture it. Charpy tests can also be used to evaluate the strain rate on the fracture, as measured with changes in the pendulum mass. Typically, only brittle and somewhat ductile polymers are evaluated with Charpy tests. In addition to the fracture energy, the type of break can be visually evaluated, as in whether the break was a total fracture of the sample or whether the sample experienced fracture in only part of the sample, and severely deformed section are still connected. Elastomers are typically not evaluated with Charpy tests due to their high yield strain inhibiting the Charpy test results.

There are many properties of polymeric materials that influence their mechanical properties. As the degree of polymerization goes up, so does the polymer's strength, as a longer chains have high Van der Waals interactions and chain entanglement. Long polymers can entangle, which leads to a subsequent increase in bulk modulus. Crazes are small cracks that form in a polymer matrix, but which are stopped by small defects in the polymer matrix. These defects are typically made up of a second, low modulus polymer that is dispersed throughout the primary phase. The crazes can increase the strength and decrease the brittleness of a polymer by allowing the small cracks to absorb higher stress and strain without leading to fracture. If crazes are allowed to propagate or coalesce, they can lead to cavitation and fracture in the sample. Crazes can be seen with transmission electron microscopy and scanning electron microscopy, and are typically engineered into a polymeric material during synthesis. Crosslinking, typically seen in thermoset polymers, can also increase the modulus, yield stress, and yield strength of a polymer.

Dynamic mechanical analysis is the most common technique used to characterize viscoelastic behavior common in many polymeric systems. DMA is also another important tool to understand the temperature dependence of polymers' mechanical behavior. Dynamic mechanical analysis is a characterization technique used to measure storage modulus and glass transition temperature, confirm crosslinking, determine switching temperatures in shape-memory polymers, monitor cures in thermosets, and determine molecular weight. An oscillating force is applied to a polymer sample and the sample's response is recorded. DMA documents the lag between force applied and deformation recovery in the sample. Viscoelastic samples exhibit a sinusoidal modulus called the dynamic modulus. Both energy recovered and lost are considered during each deformation and described quantitatively by the elastic modulus (E') and the loss modulus (E") respectively. The applied stress and the strain on the sample exhibit a phase difference, δ, which is measured over time. A new modulus is calculated each time stress is applied to the material, so DMA is used to study changes in modulus at various temperatures or stress frequencies.

Other techniques include viscometry, rheometry, and pendulum hardness.

POLYMER DEGRADATION

Polymer degradation is a change in the properties—tensile strength, color, shape, etc.—of a polymer or polymer-based product under the influence of one or more environmental factors such as heat, light or chemicals such as acids, alkalis and some salts. These changes are usually undesirable, such as cracking and chemical disintegration of products or, more rarely, desirable, as in biodegradation, or deliberately lowering the molecular weight of a polymer for recycling. The changes in properties are often termed "aging".

In a finished product such a change is to be prevented or delayed. Degradation can be useful for recycling/reusing the polymer waste to prevent or reduce environmental pollution. Degradation can also be induced deliberately to assist structure determination.

Polymeric molecules are very large (on the molecular scale), and their unique and useful properties are mainly a result of their size. Any loss in chain length lowers tensile strength and is a primary cause of premature cracking.

Commodity Polymers

Today there are primarily seven commodity polymers in use: polyethylene, polypropylene, polyvinyl chloride, polyethylene terephthalate (PET, PETE), polystyrene, polycarbonate, and poly(methyl methacrylate) (Plexiglas). These make up nearly 98% of all polymers and plastics encountered in daily life. Each of these polymers has its own characteristic modes of degradation and resistances to heat, light and chemicals. Polyethylene, polypropylene, and poly (methyl methacrylate) are sensitive to oxidation and UV radiation, while PVC may discolor at high temperatures due to loss of hydrogen chloride gas, and become very brittle. PET is sensitive to hydrolysis and attack by strong acids, while polycarbonate depolymerizes rapidly when exposed to strong alkalis.

For example, polyethylene usually degrades by *random scission*—that is by a random breakage of the linkages (bonds) that hold the atoms of the polymer together. When this polymer is heated above 450 Celsius it becomes a complex mixture of molecules of various sizes that resemble gasoline. Other polymers—like polyalphamethylstyrene—undergo 'specific' chain scission with breakage occurring only at the ends; they literally unzip or depolymerize to become the constituent monomers.

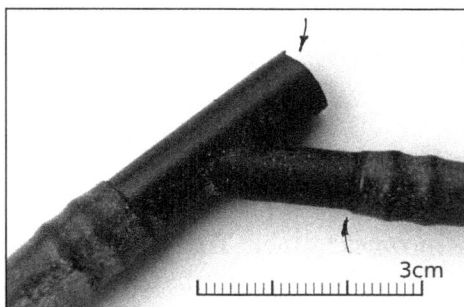

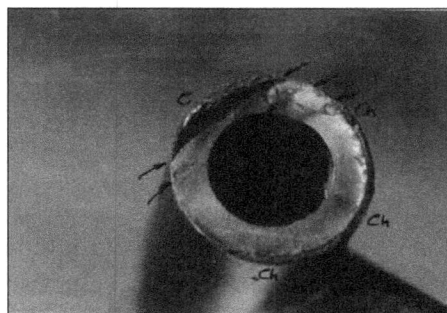

Close-up of broken fuel pipe from road traffic accident.   Close-up of broken fuel pipe connector.

Photoinduced Degradation

Most polymers can be degraded by photolysis to give lower molecular weight molecules. Electromagnetic waves with the energy of visible light or higher, such as ultraviolet light, X-rays and gamma rays are usually involved in such reactions.

Thermal Degradation

Chain-growth polymers like poly(methyl methacrylate) can be degraded by thermolysis at high temperatures to give monomers, oils, gases and water.

Thermolysis type	Added material	Temperature	Pressure	Final product
Pyrolysis		Around 500 °C	Reduced pressure	
Hydrogenation	Dihydrogen	Around 450 °C	Around 200 bars	
Gasification	Dioxygen and/or water		Under pressure	Carbon monoxide, Carbon dioxide and hydrogen

Chemical Degradation

Solvolysis

Step-growth polymers like polyesters, polyamides and polycarbonates can be degraded by solvolysis and mainly hydrolysis to give lower molecular weight molecules. The hydrolysis takes place in the presence of water containing an acid or a base as catalyst. Polyamide is sensitive to degradation by acids and polyamide mouldings will crack when attacked by strong acids. For example, the fracture surface of a fuel connector showed the progressive growth of the crack from acid attack (Ch) to the final cusp (C) of polymer. The problem is known as stress corrosion cracking, and in this case was caused by hydrolysis of the polymer. It was the reverse reaction of the synthesis of the polymer:

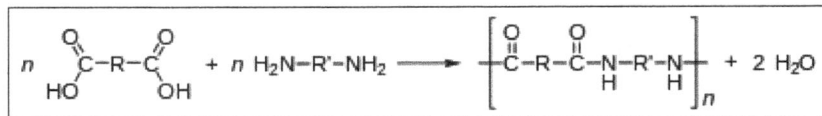

Ozonolysis

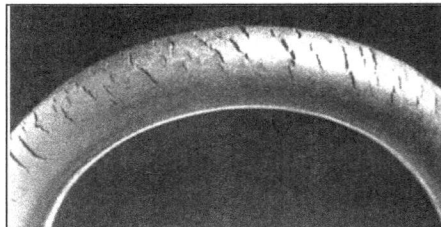
Ozone cracking in Natural rubber tubing

Cracks can be formed in many different elastomers by ozone attack. Tiny traces of the gas in the air will attack double bonds in rubber chains, with Natural rubber, polybutadiene, Styrene-butadiene rubber and NBR being most sensitive to degradation. Ozone cracks form in products under tension, but the critical strain is very small. The cracks are always oriented at right angles to the strain axis, so will form around the circumference in a rubber tube bent over. Such cracks are dangerous when they occur in fuel pipes because the cracks will grow from the outside exposed surfaces into the bore of the pipe, and fuel leakage and fire may follow. The problem of ozone cracking can be prevented by adding anti-ozonants to the rubber before vulcanization. Ozone cracks were commonly seen in automobile tire sidewalls, but are now seen rarely thanks to these additives. On the other hand, the problem does recur in unprotected products such as rubber tubing and seals.

Oxidation

The polymers are susceptible to attack by atmospheric oxygen, especially at elevated temperatures encountered during processing to shape. Many process methods such as extrusion and injection moulding involve pumping molten polymer into tools, and the high temperatures needed for melting may result in oxidation unless precautions are taken. For example, a forearm crutch suddenly snapped and the user was severely injured in the resulting fall. The crutch had fractured across a polypropylene insert within the aluminium tube of the device, and infra-red spectroscopy of the material showed that it had oxidized, possible as a result of poor moulding.

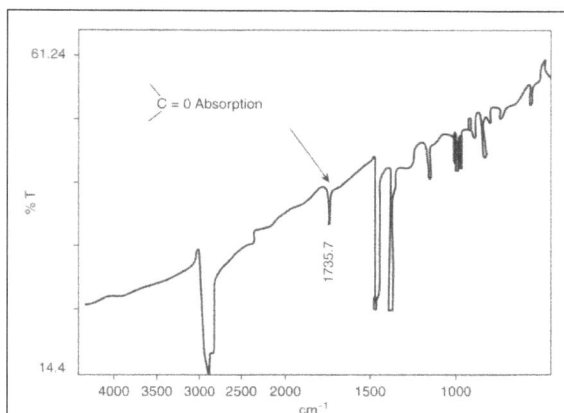

IR spectrum showing carbonyl absorption due to oxidative degradation of polypropylene crutch moulding.

Oxidation is usually relatively easy to detect owing to the strong absorption by the carbonyl group in the spectrum of polyolefins. Polypropylene has a relatively simple spectrum with few peaks at the carbonyl position (like polyethylene). Oxidation tends to start at tertiary carbon atoms because the free radicals formed here are more stable and longer lasting, making them more susceptible to attack by oxygen. The carbonyl group can be further oxidised to break the chain, this weakens the material by lowering its molecular weight, and cracks start to grow in the regions affected.

Galvanic Action

Polymer degradation by galvanic action was first described in the technical literature in 1990. This was the discovery that "plastics can corrode", i.e. polymer degradation may occur through galvanic action similar to that of metals under certain conditions and has been referred to as the "Faudree Effect". In the aerospace field, this finding has largely contributed to aircraft safety, mainly those aircraft that use CFRP and has resulted in a wide body of follow-up research and patents. Normally, when two dissimilar metals such as copper (Cu) and iron (Fe) are put into contact and then immersed in salt water, the iron will undergo corrosion, or rust. This is called a galvanic circuit where the copper is the noble metal and the iron is the active metal, i.e., the copper is the positive (+) electrode and the iron is the negative (-) electrode. A battery is formed. It follows that plastics are made stronger by impregnating them with thin carbon fibers only a few micrometers in diameter known as carbon fiber reinforced polymers (CFRP). This is to produce materials that are high strength and resistant to high temperatures. The carbon fibers act as a noble metal similar to gold (Au) or platinum (Pt). When put into contact with a more active metal, for example with aluminum (Al) in salt water the aluminum corrodes. However, in early 1990, it was reported that imide-linked resins in CFRP composites degrade when bare composite is coupled with an active metal in salt water environments. This is because corrosion not only occurs at the aluminum anode, but also at the carbon fiber cathode in the form of a very strong base with a pH of about 13. This strong base reacts with the polymer chain structure degrading the polymer. Polymers affected include bismaleimides (BMI), condensation polyimides, triazines, and blends thereof. Degradation occurs in the form of dissolved resin and loose fibers. The hydroxyl ions generated at the graphite cathode attack the O-C-N bond in the polyimide structure. Standard corrosion protection procedures were found to prevent polymer degradation under most conditions.

Chlorine-induced Cracking

Chlorine attack of acetal resin plumbing joint.

Another highly reactive gas is chlorine, which will attack susceptible polymers such as acetal resin and polybutylene pipework. There have been many examples of such pipes and acetal fittings failing in properties in the US as a result of chlorine-induced cracking. In essence, the gas attacks sensitive parts of the chain molecules (especially secondary, tertiary, or allylic carbon atoms), oxidizing the chains and ultimately causing chain cleavage. The root cause is traces of chlorine in the water supply, added for its anti-bacterial action, attack occurring even at parts per million traces of the dissolved gas. The chlorine attacks weak parts of a product, and in the case of an acetal resin junction in a water supply system, it is the thread roots that were attacked first, causing a brittle crack to grow. Discoloration on the fracture surface was caused by deposition of carbonates from the hard water supply, so the joint had been in a critical state for many months. The problems in the US also occurred to polybutylene pipework, and led to the material being removed from that market, although it is still used elsewhere in the world.

Biological Degradation

Biodegradable plastics can be biologically degraded by microorganisms to give lower molecular weight molecules. To degrade properly biodegradable polymers need to be treated like compost and not just left in a landfill site where degradation is very difficult due to the lack of oxygen and moisture.

Stabilizers

Hindered amine light stabilizers (HALS) stabilize against weathering by scavenging free radicals that are produced by photo-oxidation of the polymer matrix. UV-absorbers stabilizes against weathering by absorbing ultraviolet light and converting it into heat. Antioxidants stabilize the polymer by terminating the chain reaction due to the absorption of UV light from sunlight. The chain reaction initiated by photo-oxidation leads to cessation of crosslinking of the polymers and degradation the property of polymers.

COPOLYMER

A copolymer is a polymer that is made up of two or more monomer species. Many commercially important polymers are copolymers. Examples include polyethylene-vinyl acetate (PEVA), nitrile rubber, and acrylonitrile butadiene styrene (ABS). The process in which a copolymer is formed from multiple species of monomers is known as copolymerization. It is often used to improve or modify certain properties of plastics.

A homopolymer is a polymer that is made up of only one type of monomer unit. The difference in the constitution of a copolymer and a homopolymer is illustrated below.

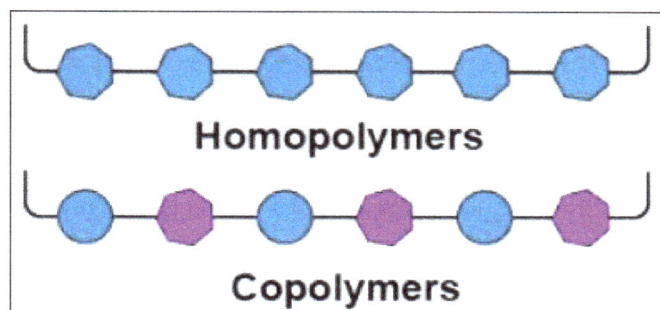

Copolymers are categorized based on their structures. Those containing a single chain are known as linear copolymers whereas those containing polymeric side chains are called branched copolymers.

Classification of Copolymers

Different Types of Linear Copolymers

Linear copolymers can be further classified into several categories such as alternating and statistical copolymers. This classification is done based on the arrangement of the monomers on the main chain.

Block Copolymers

- When more than one homopolymer units are linked together via covalent bonds, the resulting single-chain macromolecule is called a block copolymer.

- The intermediate unit at which the two homopolymer chains are linked is called a junction block.

- A diblock copolymer contains two homopolymer blocks whereas a triblock copolymer contains three distinct blocks of homopolymers.

- An example of such a polymer is acrylonitrilebutadiene styrene, commonly referred to as SBS rubber.

- An illustration describing the structure of a block copolymer which is made up of the monomers 'A' and 'B' is provided below.

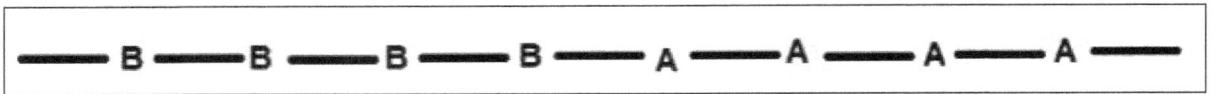

Statistical Copolymer

- Statistical copolymers are the polymers in which two or more monomers are arranged in a sequence that follows some statistical rule.
- Should the mole fraction of a monomer be equal to the probability of finding a residue of that monomer at any point in the chain, the entire polymer is then known as a random polymer.
- These polymers are generally synthesized via the free radical polymerization method.
- An example of a statistical polymer is the rubber made from the copolymers of styrene and butadiene.
- An illustration describing the structure of a statistical copolymer is provided below.

Alternating Copolymers

- Alternating copolymers contain a single main chain with alternating monomers.
- The formula of an alternating copolymer made up of monomers A and B can be generalized to $(-A-B-)_n$.
- Nylon 6,6 is an example of an alternating copolymer, consisting of alternating units of hexamethylene diamine and adipic acid.
- An illustration describing the general structure of these polymers is provided below.

Periodic Copolymers

These polymers feature a repeating sequence in which the monomers are arranged in a single chain. An illustration of the structure of a periodic copolymer made up of monomers A and B is provided below.

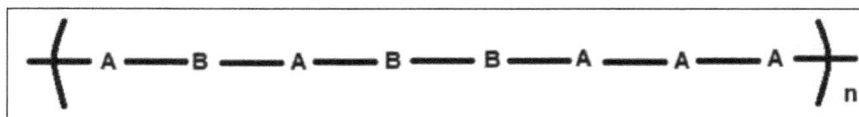

Gradient and Stereoblock Copolymers

The single-chain copolymers in which the composition of monomers gradually changes along the main chain are called gradient copolymers. If the tacticity of the monomers varies with different blocks or units in the polymer, the macromolecule is known as a stereoblock copolymer.

Branched Copolymer

As the name suggests, a branched copolymer is a polymer in which the monomers form a branched structure. Some important types of branched copolymers include star, comb, grafted, and brush copolymers.

A star copolymer contains several polymeric chains that are attached to the same central core.

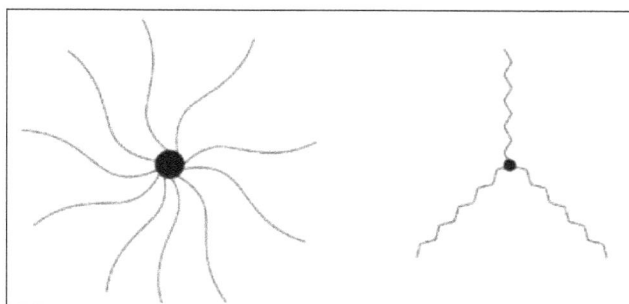

The structures of different types of star-shaped copolymers are illustrated above. They consist of a multifunctional centre to which three or more polymer chains are attached.

Graft Copolymers

Branched copolymers featuring differently structured main chains and side chains are known as graft copolymers. An illustration detailing the structure of a graft copolymer made up of monomers A and B is provided below.

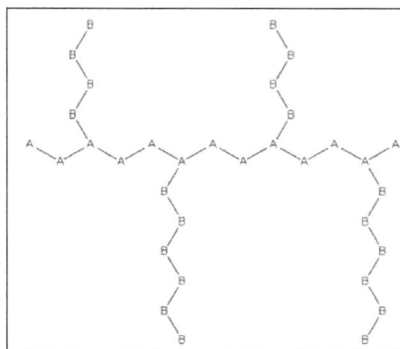

The main chain or the side chains of these polymers can be copolymers or homopolymers. High impact polystyrene is an important example of a graft copolymer. They can be synthesized from free radical polymerization.

References

- Definition-of-polymer-605912: thoughtco.com, Retrieved 5 May, 2019
- The-Basics, plastics: americanchemistry.com, Retrieved 6 June, 2019

- R. V. Lapshin; A. P. Alekhin; A. G. Kirilenko; S. L. Odintsov; V. A. Krotkov (2010). "Vacuum ultraviolet smoothing of nanometer-scale asperities of poly(methyl methacrylate) surface" (PDF). Journal of Surface Investigation. X-ray, Synchrotron and Neutron Techniques. Russia: Pleiades Publishing. 4 (1): 1–11. Doi:10.1134/S1027451010010015. ISSN 1027-4510. (Russian translation is available)

- Synthesis-of-Polymers, Organic-Chemistry-(Vollhardt-and-Schore),3A-Reactions-to-Alkenes, Organ-ic-Chemistry: chem.libretexts.org, Retrieved 7 July, 2019

- Copolymer, chemistry: byjus.com, Retrieved 8 August, 2019

5

Ceramics and Composites

A ceramic is an inorganic and non-metallic solid composed of metal, ionic, covalent or non-metal bonds. A composite material is a material that is made up of two or more constituent materials with different physical or chemical properties. The chapter closely examines the key concepts of ceramics and composites to provide an extensive understanding of the subject.

A ceramic is an inorganic, nonmetallic solid, generally based on an oxide, nitride, boride, or carbide, that is fired at a high temperature. Ceramics may be glazed prior to firing to produce a coating that reduces porosity and has a smooth, often colored surface. Many ceramics contain a mixture of ionic and covalent bonds between atoms. The resulting material may be crystalline, semi-crystalline, or vitreous. Amorphous materials with similar composition are generally termed "glass".

The four main types of ceramics are whitewares, structural ceramics, technical ceramics, and refractories. Whitewares include cookware, pottery, and wall tiles. Structural ceramics include bricks, pipes, roofing tiles, and floor tiles. Technical ceramics are also know as special, fine, advanced, or engineered ceramics. This class includes bearings, special tiles (e.g. spacecraft heat shielding), biomedical implants, ceramic brakes, nuclear fuels, ceramic engines, and ceramic coatings. Refractories are ceramics used to make crucibles, line kilns, and radiate heat in gas fireplaces.

How Ceramics are Made

Raw materials for ceramics include clay, kaolinate, aluminum oxide, silicon carbide, tungsten carbide, and certain pure elements. The raw materials are combined with water to form a mix that can be shaped or molded. Ceramics are difficult to work after they are made, so usually they are shaped into their final desired forms. The form is allowed to dry and is fired in an oven called a kiln. The firing process supplies the energy to form new chemical bonds in the material (vitrification) and sometimes new minerals (e.g., mullite forms from kaolin in the firing of porcelain). Waterproof, decorative, or functional glazes may be added prior to the first firing or may require a subsequent firing (more common). The first firing of a ceramic yields a product called the bisque. The first firing burns off organics and other volatile impurities. The second (or third) firing may be called glazing.

Examples and uses of Ceramics

Pottery, bricks, tiles, earthenware, china, and porcelain are common examples of ceramics. These materials are well-known for use in building, crafting, and art. There are many other ceramic materials:

- In the past, glass was considered a ceramic, because it's an inorganic solid that is fired and treated much like a ceramic. However, because glass is an amorphous solid, glass is usually considered to be a separate material. The ordered internal structure of ceramics plays a large role in their properties.

- Solid pure silicon and carbon may be considered to be ceramics. In a strict sense, a diamond could be called a ceramic.

- Silicon carbide and tungsten carbide are technical ceramics that have high abrasion resistance, making them useful for body armor, wear plates for mining, and machine components.

- Uranium oxide (UO_2) is a ceramic used as a nuclear reactor fuel.

- Zirconia (zirconium dioxide) is used to make ceramic knife blades, gems, fuel cells, and oxygen sensors.

- Zinc oxide (ZnO) is a semiconductor.

- Boron oxide is used to make body armor.

- Bismuth strontium copper oxide and magnesium diboride (MgB_2) are superconductors.

- Steatite (magnesium silicate) is used as an electrical insulator.

- Barium titanate is used to make heating elements, capacitors, transducers, and data storage elements.

- Ceramic artifacts are useful in archaeology and paleontology because their chemical composition can be used to identify their origin. This includes not only the composition of clay, but also that of the temper - the materials added during production and drying.

PROPERTIES OF CERAMICS

The physical properties of any ceramic substance are a direct result of its crystalline structure and chemical composition. Solid-state chemistry reveals the fundamental connection between microstructure and properties such as localized density variations, grain size distribution, type of porosity and second-phase content, which can all be correlated with ceramic properties such as mechanical strength σ by the Hall-Petch equation, hardness, toughness, dielectric constant, and the optical properties exhibited by transparent materials.

Ceramography is the art and science of preparation, examination and evaluation of ceramic microstructures. Evaluation and characterization of ceramic microstructures is often implemented on similar spatial scales to that used commonly in the emerging field of nanotechnology: from tens of angstroms (A) to tens of micrometers (μm). This is typically somewhere between the minimum wavelength of visible light and the resolution limit of the naked eye.

The microstructure includes most grains, secondary phases, grain boundaries, pores, micro-cracks, structural defects and hardness microindentions. Most bulk mechanical, optical, thermal, electrical and magnetic properties are significantly affected by the observed microstructure.

The fabrication method and process conditions are generally indicated by the microstructure. The root cause of many ceramic failures is evident in the cleaved and polished microstructure. Physical properties which constitute the field of materials science and engineering include the following:

Mechanical Properties

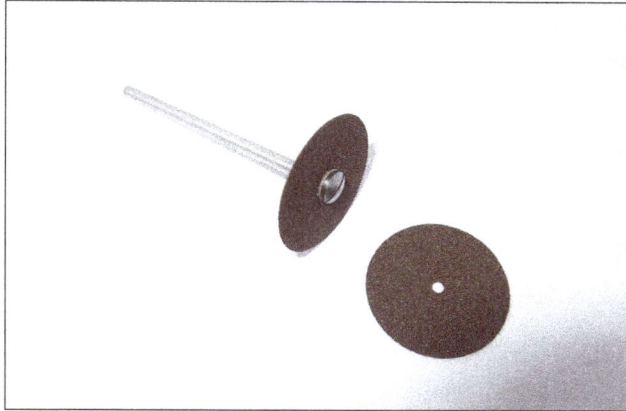
Cutting disks made of silicon carbide.

Mechanical properties are important in structural and building materials as well as textile fabrics. In modern materials science, fracture mechanics is an important tool in improving the mechanical performance of materials and components. It applies the physics of stress and strain, in particular the theories of elasticity and plasticity, to the microscopic crystallographic defects found in real materials in order to predict the macroscopic mechanical failure of bodies. Fractography is widely used with fracture mechanics to understand the causes of failures and also verify the theoretical failure predictions with real life failures.

The Porsche Carrera GT's carbon-ceramic (silicon carbide) disc brake.

Ceramic materials are usually ionic or covalent bonded materials, and can be crystalline or amorphous. A material held together by either type of bond will tend to fracture before any plastic deformation takes place, which results in poor toughness in these materials. Additionally, because these materials tend to be porous, the pores and other microscopic imperfections act as stress concentrators, decreasing the toughness further, and reducing the tensile strength. These combine to give catastrophic failures, as opposed to the more ductile failure modes of metals.

These materials do show plastic deformation. However, because of the rigid structure of the crystalline materials, there are very few available slip systems for dislocations to move, and so they deform very slowly. With the non-crystalline (glassy) materials, viscous flow is the dominant source of plastic deformation, and is also very slow. It is therefore neglected in many applications of ceramic materials.

To overcome the brittle behaviour, ceramic material development has introduced the class of ceramic matrix composite materials, in which ceramic fibers are embedded and with specific coatings are forming fiber bridges across any crack. This mechanism substantially increases the fracture toughness of such ceramics. Ceramic disc brakes are an example of using a ceramic matrix composite material manufactured with a specific process.

Ice-templating for Enhanced Mechanical Properties

Often times if a ceramic will be subjected to substantial mechanical loading it will undergo a process called Ice-templating. This process allows the finite control of the microstructure of the ceramic material and therefore the control of the mechanical properties. Ceramic engineers use this technique to tune the mechanical properties to their desired application. Specifically, Strength is increased when this technique is employed. Ice templating allows one to create macroscopic pores in a unidirectional arrangement. The applications of this oxide strengthening technique are important for solid oxide fuel cells and also water filtration devices.

In order to process a sample through ice templating a few steps are required to be completed by the ceramicist. First an aqueous colloidal suspension must be prepared containing the dissolved ceramic powder, say Yttria Stablized Zirconia (YSZ). After the ceramic precursor is ensured to be evenly dispersed throughout the colloidal solution then next processing step may begin. At this point we have a pure solution containing our YSZ dissolved powder and aqueous water in the liquid state. The solution is then cooled on a platform that allows for unidirectional cooling like the one shown in the animation to the right. The solution sample of YSZ is cooled from the bottom to the top in a unidirectional fashion. This forces ice crystals to grow in compliance to the unidirectional cooling. Inside the solution, as it is cooling, these ice crystals force the dissolved YSZ particles to the solidification front of the solid-liquid interphase boundary. At this stage of the process we have pure ice crystals lined up in a unidirectional fashion alongside concentrated pockets of the YSZ colloidal particles. The next step of the process is the sublimation step. The sample is simultaneously heated and the pressure is reduced enough to force to ice crystals to subliminate, since the sample is also heated the YSZ pockets begin to anneal together too form the first macroscopically aligned ceramic microstuctures. The sample is then further sintered to confirm the evaporation of the residual water and the final consolidation of the ceramic microstructure.

During the execution of this technique a few variables can be controlled to influence the pore size and morphology of the microstructure. These important variables of the ice-templating technique are namely the initial solids loading of the colloid, the cooling rate, the sintering temperature and time length, and also it has been shown that certain additives can influence the micro-structural morphology during this process. A good understanding of these parameters is essential to understanding the relationships between processing, microstructure, and mechanical properties of anisotropically porous materials.

Electrical Properties

Semiconductors

Some ceramics are semiconductors. Most of these are transition metal oxides that are II-VI semiconductors, such as zinc oxide.

While there are prospects of mass-producing blue LEDs from zinc oxide, ceramicists are most interested in the electrical properties that show grain boundary effects.

One of the most widely used of these is the varistor. These are devices that exhibit the property that resistance drops sharply at a certain threshold voltage. Once the voltage across the device reaches the threshold, there is a breakdown of the electrical structure in the vicinity of the grain boundaries, which results in its electrical resistance dropping from several megohms down to a few hundred ohms. The major advantage of these is that they can dissipate a lot of energy, and they self-reset – after the voltage across the device drops below the threshold, its resistance returns to being high.

This makes them ideal for surge-protection applications; as there is control over the threshold voltage and energy tolerance, they find use in all sorts of applications. The best demonstration of their ability can be found in electrical substations, where they are employed to protect the infrastructure from lightning strikes. They have rapid response, are low maintenance, and do not appreciably degrade from use, making them virtually ideal devices for this application.

Semiconducting ceramics are also employed as gas sensors. When various gases are passed over a polycrystalline ceramic, its electrical resistance changes. With tuning to the possible gas mixtures, very inexpensive devices can be produced.

Superconductivity

The Meissner effect demonstrated by levitating a magnet above a
cuprate superconductor, which is cooled by liquid nitrogen.

Under some conditions, such as extremely low temperature, some ceramics exhibit high-temperature superconductivity. The reason for this is not understood, but there are two major families of superconducting ceramics.

Ferroelectricity and Supersets

Piezoelectricity, a link between electrical and mechanical response, is exhibited by a large number of ceramic materials, including the quartz used to measure time in watches and other electronics. Such devices use both properties of piezoelectrics, using electricity to produce a mechanical motion (powering the device) and then using this mechanical motion to produce electricity (generating a signal). The unit of time measured is the natural interval required for electricity to be converted into mechanical energy and back again.

The piezoelectric effect is generally stronger in materials that also exhibit pyroelectricity, and all pyroelectric materials are also piezoelectric. These materials can be used to inter convert between thermal, mechanical, or electrical energy; for instance, after synthesis in a furnace, a pyroelectric crystal allowed to cool under no applied stress generally builds up a static charge of thousands of volts. Such materials are used in motion sensors, where the tiny rise in temperature from a warm body entering the room is enough to produce a measurable voltage in the crystal.

In turn, pyroelectricity is seen most strongly in materials which also display the ferroelectric effect, in which a stable electric dipole can be oriented or reversed by applying an electrostatic field. Pyroelectricity is also a necessary consequence of ferroelectricity. This can be used to store information in ferroelectric capacitors, elements of ferroelectric RAM.

The most common such materials are lead zirconate titanate and barium titanate. Aside from the uses mentioned above, their strong piezoelectric response is exploited in the design of high-frequency loudspeakers, transducers for sonar, and actuators for atomic force and scanning tunneling microscopes.

Positive Thermal Coefficient

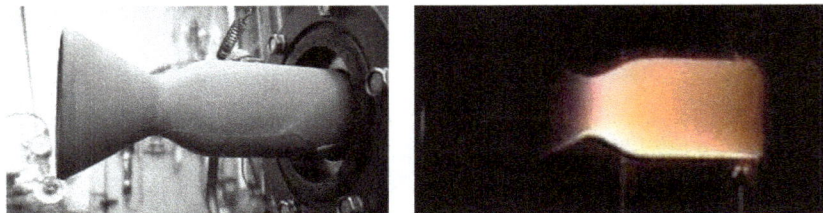

Silicon nitride rocket thruster. Left: Mounted in test stand. Right: Being tested with H_2/O_2 propellants.

Increases in temperature can cause grain boundaries to suddenly become insulating in some semiconducting ceramic materials, mostly mixtures of heavy metal titanates. The critical transition temperature can be adjusted over a wide range by variations in chemistry. In such materials, current will pass through the material until joule heating brings it to the transition temperature, at which point the circuit will be broken and current flow will cease. Such ceramics are used as self-controlled heating elements in, for example, the rear-window defrost circuits of automobiles.

At the transition temperature, the material's dielectric response becomes theoretically infinite. While a lack of temperature control would rule out any practical use of the material near its critical temperature, the dielectric effect remains exceptionally strong even at much higher temperatures. Titanates with critical temperatures far below room temperature have become synonymous with "ceramic" in the context of ceramic capacitors for just this reason.

Optical Properties

Cermax xenon arc lamp with synthetic sapphire output window.

Optically transparent materials focus on the response of a material to incoming lightwaves of a range of wavelengths. Frequency selective optical filters can be utilized to alter or enhance the brightness and contrast of a digital image. Guided lightwave transmission via frequency selective waveguides involves the emerging field of fiber optics and the ability of certain glassy compositions as a transmission medium for a range of frequencies simultaneously (multi-mode optical fiber) with little or no interference between competing wavelengths or frequencies. This resonant mode of energy and data transmission via electromagnetic (light) wave propagation, though low powered, is virtually lossless. Optical waveguides are used as components in Integrated optical circuits (e.g. light-emitting diodes, LEDs) or as the transmission medium in local and long haul optical communication systems. Also of value to the emerging materials scientist is the sensitivity of materials to radiation in the thermal infrared (IR) portion of the electromagnetic spectrum. This heat-seeking ability is responsible for such diverse optical phenomena as Night-vision and IR luminescence.

Thus, there is an increasing need in the military sector for high-strength, robust materials which have the capability to transmit light (electromagnetic waves) in the visible (0.4 – 0.7 micrometers) and mid-infrared (1 – 5 micrometers) regions of the spectrum. These materials are needed for applications requiring transparent armor, including next-generation high-speed missiles and pods, as well as protection against improvised explosive devices (IED).

In the 1960s, scientists at General Electric (GE) discovered that under the right manufacturing conditions, some ceramics, especially aluminium oxide (alumina), could be made translucent. These translucent materials were transparent enough to be used for containing the electrical plasma generated in high-pressure sodium street lamps. During the past two decades, additional types of transparent ceramics have been developed for applications such as nose cones for heat-seeking missiles, windows for fighter aircraft, and scintillation counters for computed tomography scanners.

In the early 1970s, Thomas Soules pioneered computer modeling of light transmission through translucent ceramic alumina. His model showed that microscopic pores in ceramic, mainly trapped at the junctions of microcrystalline grains, caused light to scatter and prevented true transparency. The volume fraction of these microscopic pores had to be less than 1% for high-quality optical transmission.

This is basically a particle size effect. Opacity results from the incoherent scattering of light at surfaces and interfaces. In addition to pores, most of the interfaces in a typical metal or ceramic object are in the form of grain boundaries which separate tiny regions of crystalline order. When the size of the scattering center (or grain boundary) is reduced below the size of the wavelength of the light being scattered, the scattering no longer occurs to any significant extent.

In the formation of polycrystalline materials (metals and ceramics) the size of the crystalline grains is determined largely by the size of the crystalline particles present in the raw material during formation (or pressing) of the object. Moreover, the size of the grain boundaries scales directly with particle size. Thus a reduction of the original particle size below the wavelength of visible light (~ 0.5 micrometers for shortwave violet) eliminates any light scattering, resulting in a transparent material.

Recently, Japanese scientists have developed techniques to produce ceramic parts that rival the transparency of traditional crystals (grown from a single seed) and exceed the fracture toughness of a single crystal. In particular, scientists at the Japanese firm Konoshima Ltd., a producer of ceramic construction materials and industrial chemicals, have been looking for markets for their transparent ceramics.

Livermore researchers realized that these ceramics might greatly benefit high-powered lasers used in the National Ignition Facility (NIF) Programs Directorate. In particular, a Livermore research team began to acquire advanced transparent ceramics from Konoshima to determine if they could meet the optical requirements needed for Livermore's Solid-State Heat Capacity Laser (SSHCL). Livermore researchers have also been testing applications of these materials for applications such as advanced drivers for laser-driven fusion power plants.

APPLICATIONS OF CERAMICS

Ceramics are used in an array of applications:

- Compressive strength makes ceramics good structural materials (e.g., bricks in houses, stone blocks in the pyramids).

- High voltage insulators and spark plugs are made from ceramics due to its electrical conductivity properties Good thermal insulation has ceramic tiles used in ovens and as exterior tiles on the Shuttle orbiter Some ceramics are transparent to radar and other electromagnetic waves and are used in radomes and transmitters.

- Hardness, abrasion resistance, imperviousness to high temperatures and extremely caustic conditions allow ceramics to be used in special applications where no other material can be used.

- Chemical inertness makes ceramics ideal for biomedical applications like orthopaedic prostheses and dental implants.

- Glass-ceramics, due to their high temperature capabilities, leads to uses in optical equipment and fiber insulation.

Metals, plastics & ceramics are the significant engineering material. Among the three ceramic are largely synthetic. Ceramics comprise of routine materials like cement, glass, porcelain, and brick and strange material like spacecraft & electronics. Due to the excellent power of resistance to heat and chemicals, ceramics finds wide use in industries.

Common ceramics are made from minerals such as feldspar, talc. clay, and silica, These minerals known as silicates form the majority of the earth's crust. In the laboratory chemists formulate advanced ceramics like alumina, silicon carbide, and barium titanate from mix excluding silicates.

Ceramic products resist to high temperatures, gases, water, salts and acids based on their mineral component, Properties of all ceramic products are not identical they vary from each other. Ceramics are normally bad conductor of electricity; in certain cases when cooled, they turn into super conductor. Few of the ceramics are magnetic engineers have power over these properties by scheming the ratio and nature of material being used.

Ceramics property makes them particularly appropriate products. Products made out of ceramic materials consist of construction materials, grinding materials, electrical equipment, dinnerware, glass products, and heat-resistant materials.

Ceramic materials that contain alumina and silicon carbide are extremely rigid and are used to sand various surfaces, cut metals, polish, and grind. Ceramics that consists of silica, zirconium oxide, magnesium oxide, silicon carbide & alumina are used in making refractories. Engineers continually research developing the various uses of ceramics.

Usually, Ceramic products are divided into 4 sectors:

- Structural, pipes, including bricks, roof tiles & floor.

- Refractory, such as kiln linings, gas fire radiant, steel and glass making crucibles.

- White wares, sanitary ware, including tableware, pottery products and wall tiles.

- Technical, is also known as Engineering, Advanced, Special, and Fine Ceramics. Those items include tiles used in the Space Shuttle program, ballistic protection, nuclear fuel uranium oxide pellets, gas burner nozzles, bio-medical implants, missile nose cones, and jet engine turbine blades. Repeatedly the raw materials do not include clays.

COMPOSITE MATERIAL

A composite material (also called a composition material or shortened to composite, which is the common name) is a material made from two or more constituent materials with significantly different physical or chemical properties that, when combined, produce a material with characteristics different from the individual components. The individual components remain separate and distinct within the finished structure, differentiating composites from mixtures and solid solutions.

The new material may be preferred for many reasons. Common examples include materials which are stronger, lighter, or less expensive when compared to traditional materials.

More recently, researchers have also begun to actively include sensing, actuation, computation and communication into composites, which are known as Robotic Materials.

Typical engineered composite materials include:

- Reinforced concrete and masonry.

- Composite wood such as plywood.

- Reinforced plastics, such as fibre-reinforced polymer or fiberglass.

- Ceramic matrix composites (composite ceramic and metal matrices).

- Metal matrix composites.

- and other Advanced composite materials.

Composite materials are generally used for buildings, bridges, and structures such as boat hulls, swimming pool panels, racing car bodies, shower stalls, bathtubs, storage tanks, imitation granite and cultured marble sinks and countertops.

The most advanced examples perform routinely on spacecraft and aircraft in demanding environments.

The earliest man-made composite materials were straw and mud combined to form bricks for building construction. Ancient brick-making was documented by Egyptian tomb paintings.

Wattle and daub is one of the oldest man-made composite materials, at over 6000 years old. Concrete is also a composite material, and is used more than any other man-made material in the world. As of 2006, about 7.5 billion cubic metres of concrete are made each year—more than one cubic metre for every person on Earth.

- Woody plants, both true wood from trees and such plants as palms and bamboo, yield natural composites that were used prehistorically by mankind and are still used widely in construction and scaffolding.

- Plywood 3400 BC by the Ancient Mesopotamians; gluing wood at different angles gives better properties than natural wood.

- Cartonnage layers of linen or papyrus soaked in plaster dates to the First Intermediate Period of Egypt c. 2181–2055 BC and was used for death masks.

- Cob mud bricks, or mud walls, (using mud (clay) with straw or gravel as a binder) have been used for thousands of years.

- Concrete was described by Vitruvius, writing around 25 BC in his *Ten Books on Architecture*, distinguished types of aggregate appropriate for the preparation of lime mortars. For *structural mortars*, he recommended *pozzolana*, which were volcanic sands from the sandlike beds of Pozzuoli brownish-yellow-gray in colour near Naples and reddish-brown at Rome. Vitruvius specifies a ratio of 1 part lime to 3 parts pozzolana for cements used in buildings and a 1:2 ratio of lime to pulvis Puteolanus for underwater work, essentially the same ratio mixed today for concrete used at sea. Natural cement-stones, after burning, produced cements used in concretes from post-Roman times into the 20th century, with some properties superior to manufactured Portland cement.

- Papier-mâché, a composite of paper and glue, has been used for hundreds of years.

- The first artificial fibre reinforced plastic was bakelite which dates to 1907, although natural polymers such as shellac predate it.

- One of the most common and familiar composite is fibreglass, in which small glass fibre are embedded within a polymeric material (normally an epoxy or polyester). The glass fibre is relatively strong and stiff (but also brittle), whereas the polymer is ductile (but also weak and flexible). Thus the resulting fibreglass is relatively stiff, strong, flexible, and ductile.

Composite Materials

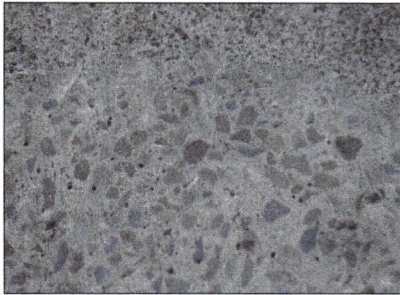

Concrete is a mixture of cement and aggregate, giving a robust, strong material that is very widely used.

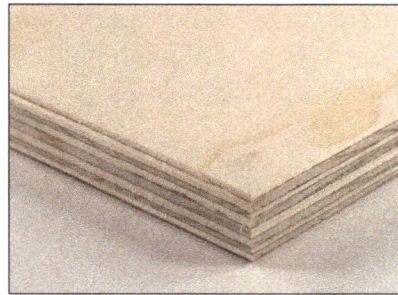

Plywood is used widely in construction.

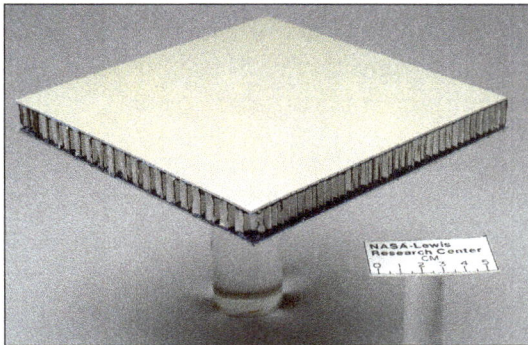

Composite sandwich structure panel used for testing at NASA.

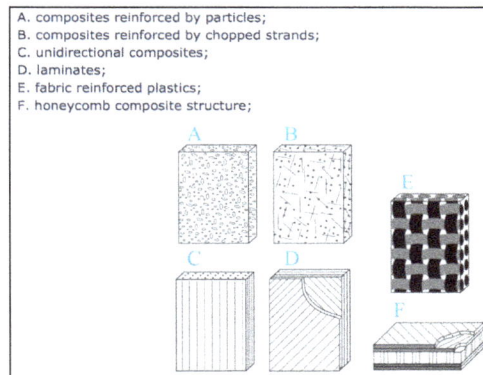

"Structural Integrity Analysis : Composites" (PDF).

Concrete is the most common artificial composite material of all and typically consists of loose stones (aggregate) held with a matrix of cement. Concrete is an inexpensive material, and will not compress or shatter even under quite a large compressive force. However, concrete cannot survive tensile loading(i.e., if stretched it will quickly break apart). Therefore, to give concrete the ability to resist being stretched, steel bars, which can resist high stretching forces, are often added to concrete to form reinforced concrete.

Fibre-reinforced polymers (FRP)s include carbon-fibre-reinforced polymer (CFRP) and glass-reinforced plastic (GRP). If classified by matrix then there are thermoplastic composites, short fibre thermoplastics, long fibre thermoplastics or long fibre-reinforced thermoplastics. There are numerous thermoset composites, including paper composite panels. Many advanced thermoset polymer matrix systems usually incorporate aramid fibre and carbon fibre in an epoxy resin matrix.

Shape memory polymer composites are high-performance composites, formulated using fibre or fabric reinforcement and shape memory polymer resin as the matrix. Since a shape memory polymer resin is used as the matrix, these composites have the ability to be easily manipulated into various configurations when they are heated above their activation temperatures and will exhibit high strength and stiffness at lower temperatures. They can also be reheated and reshaped repeatedly without losing their material properties. These composites are ideal for applications such as lightweight, rigid, deployable structures; rapid manufacturing; and dynamic reinforcement.

High strain composites are another type of high-performance composites that are designed to perform in a high deformation setting and are often used in deployable systems where structural flexing is advantageous. Although high strain composites exhibit many similarities to shape memory polymers, their performance is generally dependent on the fibre layout as opposed to the resin content of the matrix.

Composites can also use metal fibres reinforcing other metals, as in metal matrix composites (MMC) or ceramic matrix composites (CMC), which includes bone (hydroxyapatite reinforced with collagen fibres), cermet (ceramic and metal) and concrete. Ceramic matrix composites are built primarily for fracture toughness, not for strength. Another class of composite materials involve woven fabric composite consisting of longitudinal and transverse laced yarns. Woven fabric composites are flexible as they are in form of fabric.

Organic matrix/ceramic aggregate composites include asphalt concrete, polymer concrete, mastic asphalt, mastic roller hybrid, dental composite, syntactic foam and mother of pearl. Chobham armour is a special type of composite armour used in military applications.

Additionally, thermoplastic composite materials can be formulated with specific metal powders resulting in materials with a density range from 2 g/cm³ to 11 g/cm³ (same density as lead). The most common name for this type of material is "high gravity compound" (HGC), although "lead replacement" is also used. These materials can be used in place of traditional materials such as aluminium, stainless steel, brass, bronze, copper, lead, and even tungsten in weighting, balancing (for example, modifying the centre of gravity of a tennis racquet), vibration damping, and radiation shielding applications. High density composites are an economically viable option when certain materials are deemed hazardous and are banned (such as lead) or when secondary operations costs (such as machining, finishing, or coating) are a factor.

A sandwich-structured composite is a special class of composite material that is fabricated by attaching two thin but stiff skins to a lightweight but thick core. The core material is normally low strength material, but its higher thickness provides the sandwich composite with high bending stiffness with overall low density.

Wood is a naturally occurring composite comprising cellulose fibres in a lignin and hemicellulose matrix. Engineered wood includes a wide variety of different products such as wood fibre board, plywood, oriented strand board, wood plastic composite (recycled wood fibre in polyethylene matrix), Pykrete (sawdust in ice matrix), Plastic-impregnated or laminated paper or textiles,

Arborite, Formica (plastic) and Micarta. Other engineered laminate composites, such as Mallite, use a central core of end grain balsa wood, bonded to surface skins of light alloy or GRP. These generate low-weight, high rigidity materials.

Particulate composites have particle as filler material dispersed in matrix, which may be nonmetal, such as glass, epoxy. Automobile tire is an example of particulate composite.

Advanced diamond-like carbon (DLC) coated polymer composites have been reported where the coating increases the surface hydrophobicity, hardness and wear resistance.

Products

Fibre-reinforced composite materials have gained popularity (despite their generally high cost) in high-performance products that need to be lightweight, yet strong enough to take harsh loading conditions such as aerospace components (tails, wings, fuselages, propellers), boat and scull hulls, bicycle frames and racing car bodies. Other uses include fishing rods, storage tanks, swimming pool panels, and baseball bats. The Boeing 787 and Airbus A350 structures including the wings and fuselage are composed largely of composites. Composite materials are also becoming more common in the realm of orthopedic surgery, and it is the most common hockey stick material.

Carbon composite is a key material in today's launch vehicles and heat shields for the re-entry phase of spacecraft. It is widely used in solar panel substrates, antenna reflectors and yokes of spacecraft. It is also used in payload adapters, inter-stage structures and heat shields of launch vehicles. Furthermore, disk brake systems of airplanes and racing cars are using carbon/carbon material, and the composite material with carbon fibres and silicon carbide matrix has been introduced in luxury vehicles and sports cars.

In 2006, a fibre-reinforced composite pool panel was introduced for in-ground swimming pools, residential as well as commercial, as a non-corrosive alternative to galvanized steel.

In 2007, an all-composite military Humvee was introduced by TPI Composites Inc and Armor Holdings Inc, the first all-composite military vehicle. By using composites the vehicle is lighter, allowing higher payloads. In 2008, carbon fibre and DuPont Kevlar (five times stronger than steel) were combined with enhanced thermoset resins to make military transit cases by ECS Composites creating 30-percent lighter cases with high strength.

Pipes and fittings for various purpose like transportation of potable water, fire-fighting, irrigation, seawater, desalinated water, chemical and industrial waste, and sewage are now manufactured in glass reinforced plastics.

Composite materials used in tensile structures for facade application provides the advantage of being translucent. The woven base cloth combined with the appropriate coating allows better light transmission. This provides a very comfortable level of illumination compared to the full brightness of outside.

The wings of wind turbines, in growing sizes in the order of 50 m length are fabricated in composites since several years.

Two-lower-leg-amputees run on carbon-composite spring-like artifical feet as quick as healthy sportsmen.

High pressure gas cylinders typically about 7–9 litre volume x 300 bar pressure for firemen are nowadays constructed from carbon composite. Type-4-cylinders include metal only as boss that carries the thread to screw in the valve.

Carbon fibre composite part.

Composites are made up of individual materials referred to as constituent materials. There are two main categories of constituent materials: matrix (binder) and reinforcement. At least one portion of each type is required. The matrix material surrounds and supports the reinforcement materials by maintaining their relative positions. The reinforcements impart their special mechanical and physical properties to enhance the matrix properties. A synergism produces material properties unavailable from the individual constituent materials, while the wide variety of matrix and strengthening materials allows the designer of the product or structure to choose an optimum combination.

Engineered composite materials must be formed to shape. The matrix material can be introduced to the reinforcement before or after the reinforcement material is placed into the mould cavity or onto the mould surface. The matrix material experiences a melding event, after which the part shape is essentially set. Depending upon the nature of the matrix material, this melding event can occur in various ways such as chemical polymerization for a thermoset polymer matrix, or solidification from the melted state for a thermoplastic polymer matrix composite.

A variety of moulding methods can be used according to the end-item design requirements. The principal factors impacting the methodology are the natures of the chosen matrix and reinforcement materials. Another important factor is the gross quantity of material to be produced. Large quantities can be used to justify high capital expenditures for rapid and automated manufacturing technology. Small production quantities are accommodated with lower capital expenditures but higher labour and tooling costs at a correspondingly slower rate.

Many commercially produced composites use a polymer matrix material often called a resin solution. There are many different polymers available depending upon the starting raw ingredients. There are several broad categories, each with numerous variations. The most common are known as polyester, vinyl ester, epoxy, phenolic, polyimide, polyamide, polypropylene, PEEK, and others. The reinforcement materials are often fibres but also commonly ground minerals. As a rule of thumb, lay up results in a product containing 60% resin and 40% fibre, whereas vacuum infusion gives a final product with 40% resin and 60% fibre content. The strength of the product is greatly dependent on this ratio.

Martin Hubbe and Lucian A Lucia consider wood to be a natural composite of cellulose fibres in a matrix of lignin.

Constituents

Matrices

Polymers are common matrices (especially used for fibre reinforced plastics). Road surfaces are often made from asphalt concrete which uses bitumen as a matrix. Mud (wattle and daub) has seen extensive use. Typically, most common polymer-based composite materials, including fibreglass, carbon fibre, and Kevlar, include at least two parts, the substrate and the resin.

Polyester resin tends to have yellowish tint, and is suitable for most backyard projects. Its weaknesses are that it is UV sensitive and can tend to degrade over time, and thus generally is also coated to help preserve it. It is often used in the making of surfboards and for marine applications. Its hardener is a peroxide, often MEKP (methyl ethyl ketone peroxide). When the peroxide is mixed with the resin, it decomposes to generate free radicals, which initiate the curing reaction. Hardeners in these systems are commonly called catalysts, but since they do not re-appear unchanged at the end of the reaction, they do not fit the strictest chemical definition of a catalyst.

Vinyl ester resin tends to have a purplish to bluish to greenish tint. This resin has lower viscosity than polyester resin and is more transparent. This resin is often billed as being fuel resistant, but will melt in contact with gasoline. It tends to be more resistant over time to degradation than polyester resin and is more flexible. It uses the same hardeners as polyester resin (at a similar mix ratio) and the cost is approximately the same.

Epoxy resin is almost transparent when cured. In the aerospace industry, epoxy is used as a structural matrix material or as a structural glue.

Shape memory polymer (SMP) resins have varying visual characteristics depending on their formulation. These resins may be epoxy-based, which can be used for auto body and outdoor equipment repairs; cyanate-ester-based, which are used in space applications; and acrylate-based, which can be used in very cold temperature applications, such as for sensors that indicate whether perishable goods have warmed above a certain maximum temperature. These resins are unique in that their shape can be repeatedly changed by heating above their glass transition temperature (T_g). When heated, they become flexible and elastic, allowing for easy configuration. Once they are cooled, they will maintain their new shape. The resins will return to their original shapes when they are reheated above their T_g. The advantage of shape memory polymer resins is that they can be shaped and reshaped repeatedly without losing their material properties. These resins can be used in fabricating shape memory composites.

Traditional materials such as glues, muds have traditionally been used as matrices for papier-mâché and adobe.

Inorganic

Cement (concrete), metals, ceramics, and sometimes glasses are employed. Unusual matrices such as ice are sometime proposed as in pykecrete.

Reinforcements

Fiber

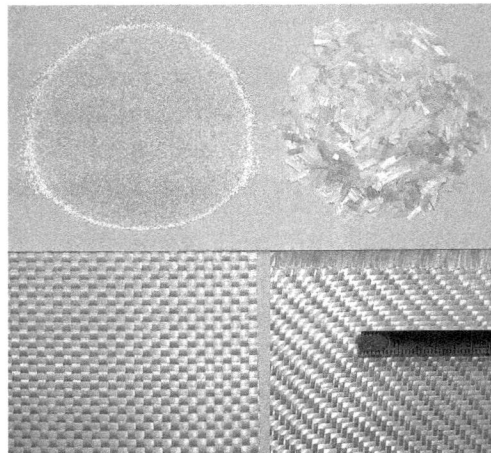

Differences in the way the fibres are laid out give different strengths and ease of manufacture.

Reinforcement usually adds rigidity and greatly impedes crack propagation. Thin fibers can have very high strength, and provided they are mechanically well attached to the matrix they can greatly improve the composite's overall properties.

Fibre-reinforced composite materials can be divided into two main categories normally referred to as short fibre-reinforced materials and continuous fiber-reinforced materials. Continuous reinforced materials will often constitute a layered or laminated structure. The woven and continuous fiber styles are typically available in a variety of forms, being pre-impregnated with the given matrix (resin), dry, uni-directional tapes of various widths, plain weave, harness satins, braided, and stitched.

The short and long fibres are typically employed in compression moulding and sheet moulding operations. These come in the form of flakes, chips, and random mate (which can also be made from a continuous fibre laid in random fashion until the desired thickness of the ply/laminate is achieved).

Common fibres used for reinforcement include glass fibres, carbon fibres, cellulose (wood/paper fibre and straw) and high strength polymers for example aramid. Silicon carbide fibers are used for some high temperature applications.

Particle

Particle reinforcement adds a similar effect to precipitation hardening in metals and ceramics. Large particles impede dislocation movement and crack propagation as well as contribute to the composite's Young's Modulus. In general, particle reinforcement effect on Young's Modulus lies between values predicted by:

$$E_c = \frac{E_\alpha E_\beta}{(V_\alpha E_\beta + V_\beta E_\alpha)}$$

as a lower bound and:

$$E_c = V_\alpha E_\alpha + V_\beta E_\beta$$

as an upper bound.

Therefore it can be expressed as a linear combination of contribution from the matrix and some weighted contribution from the particles:

$$E_c = V_m E_m + K_c V_p E_p$$

Where K_c is an experimentally derived constant between 0 and 1. This range of values for K_c reflects that particle reinforced composites are not characterized by the isostrain condition.

Similarly, the tensile strength can be modeled in an equation of similar construction where K_s is a similarly bounded constant not necessarily of the same value of K_c:

$$(T.S.)_c = V_m (T.S.)_m + K_s V_p (T.S.)_p$$

The true value of K_c and K_s vary based on factors including particle shape, particle distribution, and particle/matrix interface. Knowing these parameters, the mechanical properties can be modeled based on effects from grain boundary strengthening, dislocation strengthening, and Orowan strengthening.

The most common particle reinforced composite is concrete, which is a mixture of gravel and sand usually strengthened by addition of small rocks or sand. Metals are often reinforced with ceramics to increase strength at the cost of ductility. Finally polymers and rubber are often reinforced with carbon black, commonly used in auto tires.

Cores

Many composite layup designs also include a co-curing or post-curing of the prepreg with various other media, such as honeycomb or foam. This is commonly called a sandwich structure. This is a more common layup for the manufacture of radomes, doors, cowlings, or non-structural parts.

Open- and closed-cell-structured foams like polyvinylchloride, polyurethane, polyethylene or polystyrene foams, balsa wood, syntactic foams, and honeycombs are commonly used core materials. Open- and closed-cell metal foam can also be used as core materials. Recently, 3D graphene structures (also called graphene foam) have also been employed as core structures.

A recent review by Khurram and Xu et al., have provided the summary of the state-of-the-art techniques for fabrication of the 3D structure of graphene, and the examples of the use of these foam like structures as a core for their respective polymer composites.

Semi-crystalline Polymers

Although the two phases are chemically equivalent, semi-crystalline polymers can be described both quantitatively and qualitatively as composite materials. The crystalline portion has a higher elastic modulus and provides reinforcement for the less stiff, amorphous phase. Polymeric materials can range from 0% to 100% crystallinity aka volume fraction depending on molecular structure and thermal history. Different processing techniques can be employed to vary the percent crystallinity in these materials and thus the mechanical properties of these materials as described in the physical properties section. This effect is seen in a variety of places from industrial plastics like polyethylene shopping bags to spiders which can produce silks with different mechanical properties. In many cases these materials act like particle composites with randomly dispersed crystals known as spherulites. However they can also be engineered to be anisotropic and act more like fiber reinforced composites. In the case of spider silk, the properties of the material can even be dependent on the size of the crystals, independent of the volume fraction. Ironically, single component polymeric materials are some of the most easily tunable composite materials known.

Physical Properties

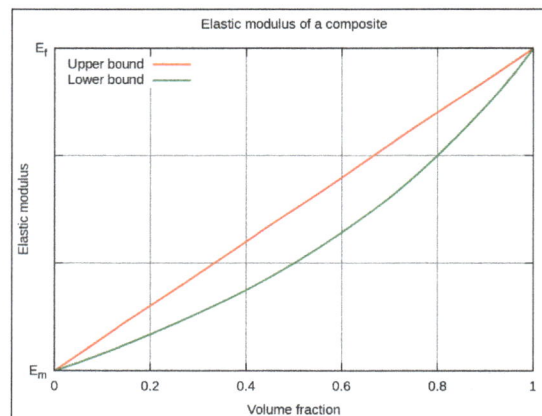

Plot of the overall strength of a composite material as a function of fiber volume fraction limited by the upper bound (isostrain) and lower bound (isostress) conditions.

The physical properties of composite materials are generally not isotropic (independent of direction of applied force) in nature, but they are typically anisotropic (different depending on the direction of the applied force or load). For instance, the stiffness of a composite panel will often depend upon the orientation of the applied forces and/or moments. The strength of a composite is bounded by two loading conditions. If both the fibres and matrix are aligned parallel to the loading direction, the deformation of both phases will be the same (assuming there is no delamination at the fibre-matrix interface). This isostrain condition provides the upper bound for composite strength, and is determined by the rule of mixtures:

$$E_C = \sum_{i=1} V_i E_i$$

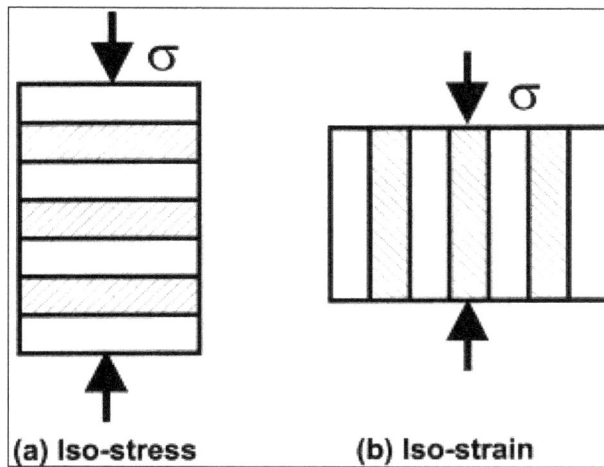

a) shows the isostress condition where the composite materials are perpendicular
to the applied force and b) is the isostrain condition that has the layers parallel to the force.

where E_c is the effective composite Young's modulus, and V_i and E_i are the volume fraction and Young's moduli, respectively, of the composite phases.

For example, a composite material made up of α and β phases under isostrain, the Young's modulus would be as follows:

$$E_C = V_\alpha E_\alpha + V_\beta E_\beta$$

where V_α and V_β are the respective volume fractions of each phase. This can be derived by considering that in the isostrain case:

$$\epsilon_C = \epsilon_\alpha = \epsilon_\beta = \epsilon$$

Assuming that the composite has a uniform cross section, the stress on the composite is a weighted average between the two phases:

$$\sigma_C = \sigma_\alpha V_\alpha + \sigma_\beta V_\beta$$

The stresses in the individual phases are given by Hooke's Law:

$$\sigma_\beta = E_\beta \epsilon$$

$$\sigma_\alpha = E_\alpha \epsilon$$

Combining these equations gives that the overall stress in the composite is:

$$\sigma_C = E_\alpha V_\alpha \epsilon + E_\beta V_\beta \epsilon = (E_\alpha V_\alpha + E_\beta V_\beta)\epsilon$$

Now we can show:

$$E_C = (E_\alpha V_\alpha + E_\beta V_\beta)$$

The lower bound is dictated by the isostress condition, in which the fibres and matrix are oriented perpendicularly to the loading direction:

$$\sigma_C = \sigma_\alpha = \sigma_\beta = \sigma$$

and now the strains become a weighted average:

$$\epsilon_C = \epsilon_\alpha V_\alpha + \epsilon_\beta V_\beta$$

Rewriting Hooke's Law for the individual phases:

$$\epsilon_\beta = \frac{\sigma}{E_\beta}$$

$$\epsilon_\alpha = \frac{\sigma}{E_\alpha}$$

This leads to:

$$\epsilon_c = V_\beta \frac{\sigma}{\epsilon_\beta} + V_\alpha \frac{\sigma}{\epsilon_\alpha} = (\frac{V_\alpha}{\epsilon_\alpha} + \frac{V_\beta}{\epsilon_\beta})\sigma$$

From the definition of Hooke's Law:

$$\frac{1}{E_C} = \frac{V_\alpha}{E_\alpha} + \frac{V_\beta}{E_\beta}$$

and in general:

$$\frac{1}{E_C} = \sum_{i=1} \frac{V_i}{E_i}$$

Following the example above, if one had a composite material made up of α and β phases under isostress conditions, the composition Young's modulus would be:

$$E_C = (E_\alpha E_\beta) / (V_\alpha E_\beta + V_\beta E_\alpha)$$

The isostrain condition implies that under an applied load, both phases experience the same strain but will feel different stress. Comparatively, under isostress conditions both phases will feel the same stress but the strains will differ between each phase. Though composite stiffness is maximized when fibres are aligned with the loading direction, so is the possibility of fibre tensile fracture, assuming the tensile strength exceeds that of the matrix. When a fibre has some angle of misorientation θ, several fracture modes are possible. For small values of θ the stress required to initiate fracture is increased by a factor of $(\cos \theta)^{-2}$ due to the increased cross-sectional area ($A \cos \theta$) of the fibre and reduced force ($F/\cos \theta$) experienced by the fibre, leading to a composite tensile strength of $\sigma_{parallel}/\cos^2 \theta$ where $\sigma_{parallel}$ is the tensile strength of the composite with fibres aligned parallel with the applied force.

Intermediate angles of misorientation θ lead to matrix shear failure. Again the cross sectional area is modified but since shear stress is now the driving force for failure the area of the matrix parallel to the fibres is of interest, increasing by a factor of 1/sin θ. Similarly, the force parallel to this area again decreases (F/cos θ) leading to a total tensile strength of τ_{my} /sinθ cosθ where τ_{my} is the matrix shear strength.

Finally, for large values of θ (near π/2) transverse matrix failure is the most likely to occur, since the fibres no longer carry the majority of the load. Still, the tensile strength will be greater than for the purely perpendicular orientation, since the force perpendicular to the fibres will decrease by a factor of 1/ sin θ and the area decreases by a factor of 1/sin θ producing a composite tensile strength of σ_{perp} /sin²θ where σ_{perp} is the tensile strength of the composite with fibres align perpendicular to the applied force.

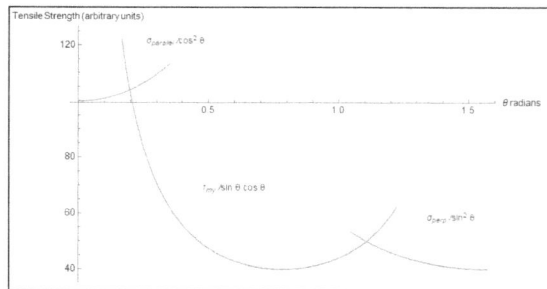

The graph depicts the three fracture modes a composite material may experience depending on the angle of misorientation relative to aligning fibres parallel to the applied stress.

The majority of commercial composites are formed with random dispersion and orientation of the strengthening fibres, in which case the composite Young's modulus will fall between the isostrain and isostress bounds. However, in applications where the strength-to-weight ratio is engineered to be as high as possible (such as in the aerospace industry), fibre alignment may be tightly controlled.

Panel stiffness is also dependent on the design of the panel. For instance, the fibre reinforcement and matrix used, the method of panel build, thermoset versus thermoplastic, and type of weave.

In contrast to composites, isotropic materials (for example, aluminium or steel), in standard wrought forms, typically have the same stiffness regardless of the directional orientation of the applied forces and/or moments. The relationship between forces/moments and strains/curvatures for an isotropic material can be described with the following material properties: Young's Modulus, the shear Modulus and the Poisson's ratio, in relatively simple mathematical relationships. For the anisotropic material, it requires the mathematics of a second order tensor and up to 21 material property constants. For the special case of orthogonal isotropy, there are three different material property constants for each of Young's Modulus, Shear Modulus and Poisson's ratio—a total of 9 constants to describe the relationship between forces/moments and strains/curvatures.

Techniques that take advantage of the anisotropic properties of the materials include mortise and tenon joints (in natural composites such as wood) and Pi Joints in synthetic composites.

Failure

Shock, impact, or repeated cyclic stresses can cause the laminate to separate at the interface between two layers, a condition known as delamination. Individual fibres can separate from the matrix e.g. fibre pull-out.

Composites can fail on the microscopic or macroscopic scale. Compression failures can occur at both the macro scale or at each individual reinforcing fibre in compression buckling. Tension failures can be net section failures of the part or degradation of the composite at a microscopic scale where one or more of the layers in the composite fail in tension of the matrix or failure of the bond between the matrix and fibres.

Some composites are brittle and have little reserve strength beyond the initial onset of failure while others may have large deformations and have reserve energy absorbing capacity past the onset of damage. The variations in fibres and matrices that are available and the mixtures that can be made with blends leave a very broad range of properties that can be designed into a composite structure. The best known failure of a brittle ceramic matrix composite occurred when the carbon-carbon composite tile on the leading edge of the wing of the Space Shuttle Columbia fractured when impacted during take-off. It led to catastrophic break-up of the vehicle when it re-entered the Earth's atmosphere on 1 February 2003.

Compared to metals, composites have relatively poor bearing strength.

Testing

To aid in predicting and preventing failures, composites are tested before and after construction. Pre-construction testing may use finite element analysis (FEA) for ply-by-ply analysis of curved surfaces and predicting wrinkling, crimping and dimpling of composites. Materials may be tested during manufacturing and after construction through several nondestructive methods including ultrasonics, thermography, shearography and X-ray radiography, and laser bond inspection for NDT of relative bond strength integrity in a localized area.

FABRICATION METHODS

There are numerous methods for fabricating composite components. Some methods have been borrowed (injection molding, for example), but many were developed to meet specific design or manufacturing challenges. Selection of a method for a particular part, therefore, will depend on the materials, the part design and end-use or application. Composite fabrication processes involve some form of molding, to shape the resin and reinforcement. A mold tool is required to give the unformed resin/fiber combination its shape prior to and during cure. The most basic fabrication method for thermoset composites is hand layup, which typically consists of laying dry fabric layers, or "plies," or prepreg plies, by hand onto a tool to form a laminate stack. Resin is applied to the dry plies after layup is complete (e.g., by means of resin infusion). Several curing methods are available. The most basic is simply to allow cure to occur at room temperature. Cure can be accelerated, however, by applying heat, typically with an oven, and pressure, by means of a vacuum.Many high-performance thermoset parts require heat and high consolidation pressure to cure — conditions that require the use of an autoclave. Autoclaves, generally, are expensive to buy and operate. Manufacturers that are equipped with autoclaves usually cure a number of parts simultaneously. Computer systems monitor and control autoclave temperature, pressure, vacuum and inert atmosphere, which allows unattended and/or remote supervision of the cure process and maximizes efficient use of the technique. Electron-beam (E-beam) curing has been explored as an efficient curing method for thin laminates. In Ebeam curing, the composite layup is exposed to a stream of electrons that provide ionizing radiation, causing polymerization and crosslinking in radiationsensitive resins. X-ray and microwave curing technologies work in a similar manner. A fourth alternative, ultraviolet (UV) curing, involves the use of UV radiation to activate a photoinitiator added to a thermoset resin, which, when activated, sets off a crosslinking reaction. UV curing requires lightpermeable resin and reinforcements.

Open Molding

Open contact molding in one-sided molds is a lowcost, common process for making fiberglass composite products. Typically used for boat hulls and decks, RV components, truck cabs and fenders, spas, bathtubs, shower stalls and other relatively large, noncomplex shapes, open molding involves either hand layup or a semi-automated alternative, sprayup. In an open-mold sprayup application, the mold is first treated with mold release. If a gel coat is used, it is typically sprayed into the mold after the mold release has been applied. The gel coat then is cured and the mold is ready for fabrication to begin. In the sprayup process, catalyzed resin (viscosity from 500 to 1,000 cps) and glass fiber are sprayed into the mold using a chopper gun, which chops continuous fiber into short lengths, then blows the short fibers directly into the sprayed resin stream so that both materials are applied simultaneously. To reduce VOCs, piston pump-activated, non-atomizing spray guns and fluid impingement spray heads dispense gel coats and resins in larger droplets at low pressure. Another option is a roller impregnator, which pumps resin into a roller similar to a paint roller. In the final steps of the sprayup process, workers compact the laminate by hand with rollers. Wood, foam or other core material may then be added, and a second sprayup layer imbeds the core between the laminate skins. The part is then cured, cooled and removed from the reusable mold. Hand layup and sprayup methods are often used in tandem to reduce labor.

Resin Infusion Processes

Ever-increasing demand for faster production rates has pressed the industry to replace hand layup with alternative fabrication processes and has encouraged fabricators to automate those processes wherever possible.

A common alternative is resin transfer molding (RTM), sometimes referred to as liquid molding. The benefits of RTM are impressive. Generally, the dry preforms and resins used in RTM are less expensive than prepreg material and can be stored at room temperature. The process can produce thick, near-net shape parts, eliminating most post-fabrication work. It also yields dimensionally accurate complex parts with good surface detail and delivers a smooth finish on all exposed surfaces. It is possible to place inserts inside the preform before the mold is closed, allowing the RTM process to accommodate core materials and integrate "molded in" fittings and other hardware into the part structure. Finally, RTM significantly cuts cycle times and can be adapted for use as one stage in an automated, repeatable manufacturing process for even greater efficiency, reducing cycle time from what can be several days, typical of hand layup, to just hours — or even minutes.

In contrast to RTM, where resin and catalyst are premixed prior to injection under pressure into the mold, reaction injection molding (RIM) injects a rapid-cure resin and a catalyst into the mold in two separate streams. Mixing and the resulting chemical reaction occur in the mold instead of in a dispensing head. Automotive industry suppliers combine structural RIM (SRIM) with rapid preforming methods to fabricate structural parts that don't require a Class A finish. Programmable robots have become a common means to spray a chopped fiberglass/binder combination onto a vacuumequipped preform screen or mold. Robotic sprayup can be directed to control fiber orientation. A related technology, dry fiber placement, combines stitched preforms and RTM. Fiber volumes of up to 68 percent are possible, and automated controls ensure low voids and consistent preform reproduction, without the need for trimming.

Vacuum-assisted resin transfer molding (VARTM) refers to a variety of related processes that represent the fastest-growing new molding technology. The salient difference between VARTM-type processes and RTM is that in VARTM, resin is drawn into a preform through use of a vacuum only, rather than pumped in under pressure. VARTM does not require high heat or pressure. For that reason, VARTM operates with low-cost tooling, making it possible to inexpensively produce large, complex parts in one shot.

In the VARTM process, fiber reinforcements are placed in a one-sided mold, and a cover (typically a plastic bagging film) is placed over the top to form a vacuum-tight seal. The resin typically enters the structure through strategically placed ports and feed lines, termed a "manifold." It is drawn by vacuum through the reinforcements by means of a series of designed-in channels that facilitate wetout of the fibers. Fiber content in the finished part can run as high as 70 percent. Current applications include marine, ground transportation and infrastructure parts. This method has been employed by The Boeing Co. (Chicago, Ill.) and NASA, as well as small fabricating firms, to produce aerospace-quality laminates without an autoclave.

High-volume Molding Methods

Compression molding is a high-volume thermoset molding process that employs expensive but very durable metal dies. It is an appropriate choice when production quantities exceed 10,000 parts. As many as 200,000 parts can be turned out on a set of forged steel dies, using sheet molding compound (SMC), a composite sheet material made by sandwiching chopped fiberglass between two layers of thick resin paste. Low-pressure SMC formulations that are now on the market offer open molders low-capitalinvestment entry into closed-mold processing with near-zero VOC emissions and the potential for very high-quality surface finish. Automakers are exploring carbon fiber-reinforced SMC, hoping to take advantage of carbon's high strength- and stiffness-to-weight ratios in exterior body panels and other parts. Newer, toughened SMC formulations help prevent microcracking, a phenomenon that previously caused paint "pops" during the painting process (surface craters caused by outgassing, the release of gasses trapped in the microcracks during oven cure). Composites manufacturers in industrial markets are formulating their own resins and compounding SMC in-house to meet needs in specific applications that require UV, impact and moisture resistance and have surface-quality demands that drive the need for customized material development.

Injection molding is a fast, high-volume, lowpressure, closed process using, most commonly, filled thermoplastics, such as nylon with chopped glass fiber. In the past 20 years, however, automated injection molding of BMC has taken over some markets previously held by thermoplastic and metal casting manufacturers. For example, the first-ever BMC-based electronic throttle control (ETC) valves (previously molded only from die-cast aluminum) debuted on engines in the BMW Mini and the Peugeot 207, taking advantage of dimensional stability offered by a specially-formulated BMC supplied by TetraDUR GmbH (Hamburg, Germany), a subsidiary of Bulk Molding Compounds Inc. (BMCI, West Chicago, Ill.,).Injection speeds are typically one to five seconds, and as many as 2,000 small parts can be produced per hour in some multiple-cavity molds.

Parts with thick cross-sections can be compression molded or transfer molded with BMC. Transfer molding is a closed-mold process wherein a measured charge of BMC is placed in a pot with runners that lead to the mold cavities. A plunger forces the material into the cavities, where the product cures under heat and pressure.

Filament winding is a continuous fabrication method that can be highly automated and repeatable, with relatively low material costs. Filament winding yields parts with exceptional circumferential or "hoop" strength. The highest-volume single application of filament winding is golf club shafts. Fishing rods, pipe, pressure vessels and other cylindrical parts comprise most of the remaining business.

Pultrusion, like RTM, has been used for decades with glass fiber and polyester resins, but in the last 10 years the process also has found application in advanced composites applications. In this relatively simple, low-cost, continuous process, the reinforcing fiber (usually roving, tow or continuous mat) is typically pulled through a heated resin bath and then formed into specific shapes as it passes through one or more forming guides or bushings. The material then moves through a heated die, where it takes its net shape and cures. Further downstream, after cooling, the resulting profile is cut to desired length. Pultrusion yields smooth finished parts that typically do not require postprocessing. A wide range of continuous, consistent, solid and hollow profiles are pultruded, and the process can be custom-tailored to fit specific applications.

Tube rolling is a longstanding composites manufacturing process that can produce finite-length tubes and rods. It is particularly applicable to smalldiameter cylindrical or tapered tubes in lengths as great as 20 ft/6.2m. Tubing diameters up to 6 inches/152 mm can be rolled efficiently. Typically, a tacky prepreg fabric or unidirectional tape is used, depending on the part. The material is precut in patterns that have been designed to achieve the requisite ply schedule and fiber architecture for the application. The pattern pieces are laid out on a flat surface and a mandrel is rolled over each one under applied pressure, which compacts and debulks the material. When rolling a tapered mandrel — e.g., for a fishing rod or golf shaft — only the first row of longitudinal fibers falls on the true 0° axis. To impart bending strength to the tube, therefore, the fibers must be continuously reoriented by repositioning the pattern pieces at regular intervals.

Automated fiber placement (AFP) The fiber placement process automatically places multiple individual prepreg tows onto a mandrel at high speed, using a numerically controlled, articulating robotic placement head to dispense, clamp, cut and restart as many as 32 tows simultaneously. Minimum cut length (the shortest tow length a machine can lay down) is the essential ply-shape determinant. The fiber placement heads can be attached to a 5-axis gantry, retrofitted to a filament winder or delivered as a turnkey custom system. Machines are available with dual mandrel stations to increase productivity. Advantages of fiber placement include processing speed, reduced material scrap and labor costs, parts consolidation and improved part-to-part uniformity. Often, the process is used to produce large thermoset parts with complex shapes.

Automated tape laying (ATL) is an even speedier automated process in which prepreg tape, rather than single tows, is laid down continuously to form parts. It is often used for parts with highly complex contours or angles. Tape layup is versatile, allowing breaks in the process and easy direction changes, and it can be adapted for both thermoset and thermoplastic materials. The head includes a spool or spools of tape, a winder, winder guides, a compaction shoe, a position sensor and a tape cutter or slitter. In either case, the head may be located on the end of a multi axis articulating robot that moves around the tool or mandrel to which material is being applied, or the head may be located on a gantry suspended above the tool. Although ATL generally is faster than AFP and can place more material over longer distances, AFP is better suited to shorter courses and can place material more effectively over contoured surfaces. These technologies grew out of the machine tool industry and have seen extensive use in the manufacture of the fuselage, wing skin panels, wingbox, tail and other structures on the forthcoming Boeing 787 Dreamliner and the Airbus A350 XWB. ATL and AFP also are used extensively to produce parts for the F-35 Lightning II fighter jet the V-22 Osprey tiltrotor troop transport and a variety of other aircraft.

Centrifugal casting of pipe from 1 inch/25 mm to 14 inches/356 mm in diameter is an alternative to filament winding for high-performance, corrosionresistant service. In cast pipe, 0°/90° woven fiberglass provides both longitudinal and hoop strength throughout the pipe wall and brings greater strength at equal wall thickness compared to multiaxial fiberglass wound pipe. In the casting process, epoxy or vinyl ester resin is injected into a 150G centrifugally spinning mold, permeating the woven fabric wrapped around the mold's interior surface. The centrifugal force pushes the resin through the layers of fabric, creating a smooth finish on the outside of the pipe, and excess resin pumped into the mold creates a resin-rich, corrosion- and abrasionresistant interior liner. Fiber-reinforced thermoplastic components now can be produced by extrusion, as well. A huge market has emerged in the past decade for extruded thermoplastic/wood flour (or other additives, such as bast fibers or fly ash) composites. These wood plastic composites, or WPCs, used to simulate wood decking, siding, window and door frames, and fencing.

References

- Ceramic-definition-chemistry: thoughtco.com, Retrieved 6 August, 2019

- Mississippi Valley Archaeological Center, Ceramic analysisarchived June 3, 2012, at the Wayback Machine, Retrieved 04-11-12

- Industrial-ceramics-products: cumi-murugappa.com, Retrieved 9 January, 2019

- Quan, Hui; Li, Zhong-Ming; Yang, Ming-Bo; Huang, Rui (June 2005). "On transcrystallinity in semi-crystalline polymer composites". Composites Science and Technology. 65 (7–8): 999–1021. Doi:10.1016/j.compscitech.2004.11.015

- Black, J. T.; Kohser, R. A. (2012). DeGarmo's materials and processes in manufacturing. Wiley. p. 226. ISBN 978-0-470-92467-9

- "What is Resin Transfer Moulding (RTM)?". Coventive Composites. 2018-04-25. Retrieved 2018-10-01

6

Nanomaterials and its Types

Nanomaterials are the materials whose size lies between 1 and 1000 nanometers. A few of the main types of nanomaterials are nanoparticles, carbon nanotube, graphene, nanocomposites and nanostructured films. This chapter has been carefully written to provide an easy understanding of these types of nanomaterials.

Nanomaterials are the particles (crystalline or amorphous) of organic or inorganic materials having sizes in the range of 1-100 nm. Nanomaterials are classified into nanostructured materials and nanophase/nanoparticle materials. The former refer to condensed bulk materials that are made of grains with grain sizes in the nanometer size range while the latter are usually the dispersive nanoparticles. To distinguish nanomaterials from bulk, it is vitally important to demonstrate the unique properties of nanomaterials and their prospective impacts in science and technology.

Technology in the twenty first century requires the miniaturization of devices in to nanometer sizes while their ultimate performance is dramatically enhanced. This raises many issues regarding to new materials for achieving specific functionality and selectivity. Nanotechnology is the design, fabrication and application of nanostructures or nanomaterials and the fundamental understanding of the relationships between physical properties or phenomena and material dimensions. It is a new field or a new scientific domain. Nanotechnology also promises the possibility of creating nanostructures of metastable phases with non–conventional properties including superconductivity and magnetism. Another very important aspect of nanotechnology is the miniaturization of current and new instruments, sensors and machines that will greatly impact the world we live in. Examples of possible miniaturization are computers with infinitely great power that compute algorithms to mimic human brains, biosensors that warn us at the early stage of the onset of disease and preferably at the molecular level and target specific drugs that automatically attack the diseased cells on site, nanorobots that can repair internal damage and remove chemical toxins in human bodies, nanoscaled electronics that constantly monitor our local environment.

Nanomaterials have properties that are significantly different and considerably improved relative to those of their coarser-grained counterparts. The property changes result from their small grain sizes, the large percentage of their atoms in large grain boundary environments and the interaction between the grains. Research on a variety of chemical, mechanical and physical properties is beginning to yield a glimmer of understanding of just how this interplay manifests itself in the properties of these new materials. In general, one can have nanoparticles of metals, semiconductors, dielectrics, magnetic materials, polymers or other organic compounds. Semiconductor heterostructures are usually referred to as one-dimensional artificially structured materials composed of layers of different phases/compositions. The semiconductor heterostructured material is the optimum candidate for fabricating electronic and photonic nanodevices.

It is seen that properties of these particles are quite sensitive to their sizes. This is partly connected with the fact that surface to volume ratio changes with a change in particle size. A high percentage of surface atoms introduce many size-dependent phenomena. High surface area is an important feature of nanosized and nanoporous materials, which can be exploited in many potential industrial applications, such as separation science and catalytic processing, because of the enhanced chemical reactivity.

For optical applications, a wide range of nanostructure-based optical sources that include high performance lasers to general illumination can be fabricated. These industrial requirements can be accomplished by selecting an appropriate fabrication method of functional nanostructures with controlled size, shape and composition. However, assembling nanoparticles to form a nanostructure is a complex process. Numerous research groups are working out different synthetic strategies to find economically affordable ways for fabricating the nanostructures and simultaneously preserving the superior characteristics of the basic building units (nanoparticles) in various devices.

NANOPARTICLES

Nanoparticle is an ultrafine unit with dimensions measured in nanometres (nm; 1 nm = 10^{-9} metre). Nanoparticles exist in the natural world and are also created as a result of human activities. Because of their submicroscopic size, they have unique material characteristics, and manufactured nanoparticles may find practical applications in a variety of areas, including medicine, engineering, catalysis, and environmental remediation.

Properties of Nanoparticles

In 2008 the International Organization for Standardization (ISO) defined a nanoparticle as a discrete nano-object where all three Cartesian dimensions are less than 100 nm. The ISO standard similarly defined two-dimensional nano-objects (i.e., nanodiscs and nanoplates) and one-dimensional nano-objects (i.e., nanofibres and nanotubes). But in 2011 the Commission of the European Union endorsed a more-technical but wider-ranging definition:

> "a natural, incidental or manufactured material containing particles, in an unbound state or as an aggregate or as an agglomerate and where, for 50% or more of the particles in the number size distribution, one or more external dimensions is in the size range 1 nm–100 nm."

Under that definition a nano-object needs only one of its characteristic dimensions to be in the range 1–100 nm to be classed as a nanoparticle, even if its other dimensions are outside that range. The lower limit of 1 nm is used because atomic bond lengths are reached at 0.1 nm.

That size range—from 1 to 100 nm—overlaps considerably with that previously assigned to the field of colloid science—from 1 to 1,000 nm—which is sometimes alternatively called the mesoscale.

Thus, it is not uncommon to find literature that refers to nanoparticles and colloidal particles in equal terms. The difference is essentially semantic for particles below 100 nm in size.

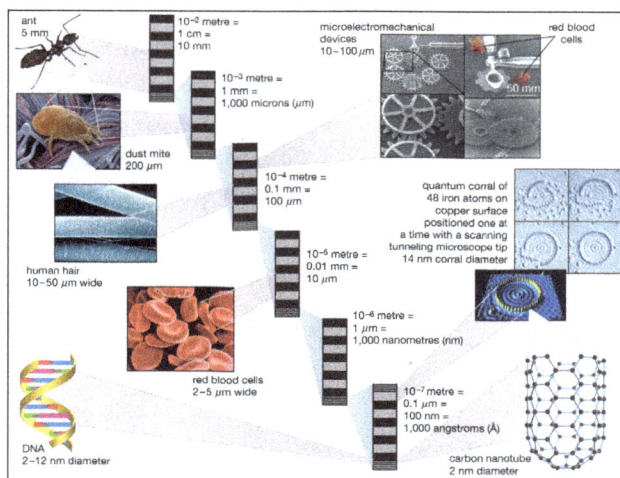

Examples from biological and mechanical realms illustrate various "orders of magnitude" (powers of 10), from 10⁻² metre down to 10⁻⁷ metre.

Nanoparticles can be classified into any of various types, according to their size, shape, and material properties. Some classifications distinguish between organic and inorganic nanoparticles; the first group includes dendrimers, liposomes, and polymeric nanoparticles, while the latter includes fullerenes, quantum dots, and gold nanoparticles. Other classifications divide nanoparticles according to whether they are carbon-based, ceramic, semiconducting, or polymeric. In addition, nanoparticles can be classified as hard (e.g., titania[titanium dioxide], silica [silica dioxide] particles, and fullerenes) or as soft (e.g., liposomes, vesicles, and nanodroplets). The way in which nanoparticles are classified typically depends on their application, such as in diagnosis or therapy versus basic research, or may be related to the way in which they were produced.

There are three major physical properties of nanoparticles, and all are interrelated: (1) they are highly mobile in the free state (e.g., in the absence of some other additional influence, a 10-nm-diameter nanosphere of silica has a sedimentation rate under gravity of 0.01 mm/day in water); (2) they have enormous specific surface areas (e.g., a standard teaspoon, or about 6 ml, of 10-nm-diameter silica nanospheres has more surface area than a dozen doubles-sized tennis courts; 20 percent of all the atoms in each nanosphere will be located at the surface); and (3) they may exhibit what are known as quantum effects. Thus, nanoparticles have a vast range of compositions, depending on the use or the product.

Nanoparticle-based Technologies

In general, nanoparticle-based technologies centre on opportunities for improving the efficiency, sustainability, and speed of already-existing processes. That is possible because, relative to the materials used traditionally for industrial processes (e.g., industrial catalysis), nanoparticle-based technologies use less material, a large proportion of which is already in a more "reactive" state. Other opportunities for nanoparticle-based technologies include the use of nanoscale zero-valent iron (NZVI) particles as a field-deployable means of remediating organochlorine compounds, such as polychlorinated biphenyls (PCBs), in the environment. NZVI particles are able to permeate into rock layers in the ground and thus can neutralize the reactivity of organochlorines in deep aquifers. Other applications of nanoparticles are those that stem from manipulating or arranging matter at the nanoscale to provide better coatings, composites, or additives and those that exploit the particles' quantum effects (e.g., quantum dots for imaging, nanowires for molecular electronics, and technologies for spintronics and molecular magnets).

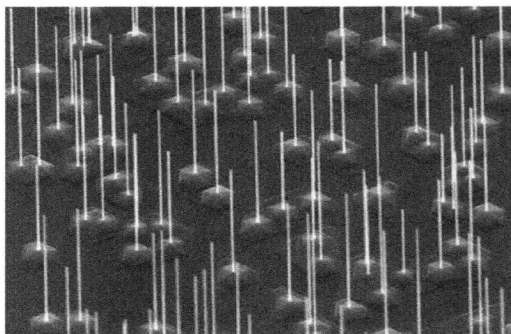

Nanowires as seen by a field-emission microscope.

Nanoparticle Applications in Materials

Many properties unique to nanoparticles are related specifically to the particles' size. It is therefore natural that efforts have been made to capture some of those properties by incorporating nanoparticles into composite materials. An example of how the unique properties of nanoparticles have been put to use in a nanocomposite material is the modern rubber tire, which typically is a composite of a rubber (an elastomer) and an inorganic filler (a reinforcing particle), such as carbon black or silica nanoparticles.

For most nanocomposite materials, the process of incorporating nanoparticles is not straightforward. Nanoparticles are notoriously prone to agglomeration, resulting in the formation of large clumps that are difficult to redisperse. In addition, nanoparticles do not always retain their unique size-related properties when they are incorporated into a composite material.

Despite the difficulties with manufacture, the use of nanomaterials grew markedly in the early 21st century, with especially rapid growth in the use of nanocomposites. Nanocomposites were employed in the development and design of new materials, serving, for example, as the building blocks for new dielectric (insulating) and magnetic materials.

Polymers

Similar to the way in which carbon and silica nanoparticles have been used as fillers in rubber to improve the mechanical properties of tires, such particles and others, including nanoclays, have been incorporated into polymers to improve their strength and impact resistance. In the early 21st century, increasing use of non-petroleum-based polymers that were derived from natural sources drove the development of "all-natural" nanocomposite polymers. Such materials incorporate a biopolymer derived from an alginate (a carbohydrate found in the cell wall of brown algae), cellulose, or starch; the biopolymer is used in conjunction with a natural nanoclay or a filler derived from the shells of crustaceans. The materials are biodegradable and do not leave behind potentially harmful or nonnatural residues.

Food Packaging

Nanoparticles have been increasingly incorporated into food packaging to control the ambient atmosphere around food, keeping it fresh and safe from microbial contamination. Such composites use nanoflakes of clays and claylike particles, which slow down the ingress of moisture and reduce gas transport across the packaging film. It is also possible to incorporate nanoparticles with apparent antimicrobial effects (e.g., nanocopper or nanosilver) into such packaging. Nanoparticles that exhibit antimicrobial activity had also been incorporated into paints and coatings, making those products particularly useful for surfaces in hospitals and other medical facilities and in areas of food preparation.

Flame Retardants

Nanoparticles were explored for their potential to replace additives based on flammable organic halogens and phosphorus in plastics and textiles. Studies had suggested that, in the event of a serious fire, products with nanoclays and hydroxide nanoparticles were associated with fewer emissions of harmful fumes than products containing certain other types of additives.

Batteries and Supercapacitors

The ability to engineer nanocomposite materials to have very high internal surface areas for storing electrical charge in the form of small ions or electrons has made them especially valuable for use in batteries and supercapacitors. Indeed, nanocomposite materials have been synthesized for various applications involving electrodes. Composite materials based on carbon nanotubes and layered-type materials, such as graphene, were also researched extensively, making their first appearances in commercial devices in the early 2000s.

Nanoceramics

A long-term objective in materials science had been to transform ceramics that are brittle and prone to cracking into tougher, more resilient materials. By the early 21st century, researchers had achieved that goal by incorporating an effective blend of nanoparticles into ceramics materials. Other new ceramics materials that were under development included all-ceramic or polymer-ceramic blends, which combined the unique functional (e.g., electrical, magnetic, or mechanical) properties of a nanocomposite material with the properties of ceramics materials.

Light Control

In the 1990s the development of blue light-emitting diodes (LEDs), which had the potential to produce white light at significantly reduced costs, inspired a revolution in lighting. Blue LEDs brought about a need for composite materials that could be used to coat the diodes to convert blue light into other wavelengths (such as red, yellow, or green) in order to achieve white light. One way of obtaining the desired light is by leveraging the size or quantum effect of small semiconducting particles. The application of such particles facilitated the development of nanocomposite polymers for greenhouse enclosures; the polymers optimize plant growth by effectively converting wavelengths of full-spectrum sunlight into the red and blue wavelengths used in photosynthesis. Light conversion in the above cases is achieved with submicron particles of inorganic phosphor materials incorporated into the polymer.

Nanoparticle Applications in Medicine

The small size of nanoparticles is especially advantageous in medicine; nanoparticles can not only circulate widely throughout the body but also enter cells or be designed to bind to specific cells. Those properties have enabled new ways of enhancing images of organs as well as tumours and other diseased tissues in the body. They also have facilitated the development of new methods of delivering therapy, such as by providing local heating (hyperthermia), by blocking vasculature to diseased tissues and tumours, or by carrying payloads of drugs.

Magnetic nanoparticles have been used to replace radioactive technetium for tracking the spread of cancer along lymph nodes. The nanoparticles work by exploiting the change in contrast brought about by tiny particles of superparamagnetic iron oxide in magnetic resonance imaging (MRI). Such particles also can be used to kill tumours via hyperthermia, in which an alternating magnetic field causes them to heat and destroy tissue on a local scale.

Nanoparticles can be designed to enhance fluorescent imaging or to enhance images from positron emission tomography (PET) or ultrasound. Those methods typically require that the nanoparticle be able to recognize a particular cell or disease state. In theory, the same idea of targeting could be used in aiding the precise delivery of a drug to a given disease site. The drug could be carried via a nanocapsule or a liposome, or it could be carried in a porous nanosponge structure and then held by bonds at the targeted site, thereby allowing the slow release of drug. The development of nanoparticles to aid in the delivery of a drug to the brainvia inhalation holds considerable promise for the treatment of neurological disorders such as Parkinson disease, Alzheimer disease, and multiple sclerosis.

Nanoparticles and nanofibres play an important part in the design and manufacture of novel scaffold structures for tissue and bone repair. The nanomaterials used in such scaffolds are biocompatible. For example, nanoparticles of calcium hydroxyapatite, a natural component of bone, used in combination with collagen or collagen substitutes could be used in future tissue-repair therapies.

Nanoparticles also have been used in the development of health-related products. For example, a sunscreen known as Optisol, invented at the University of Oxford in the 1990s, was designed with the objective of developing a safe sunscreen that was transparent in visible light but retained ultraviolet-blocking action on the skin. The ingredients traditionally used in sunscreens were based on large particles of either zinc oxide or titanium dioxide or contained an organic sunlight-absorbing compound. However, those materials were not satisfactory: zinc oxide and titanium dioxide are very potent photocatalysts, and in the presence of water and sunlight they generate free radicals, which have the potential to damage skin cells and DNA(deoxyribonucleic acid). Scientists proceeded to develop a nanoparticle form of titanium oxide that contained a small amount of manganese. Studies indicated that the nanoparticle-based sunscreen was safer than sunscreen products manufactured by using traditional materials. The improvement in safety was attributed to the introduction of manganese, which changed the semiconducting properties of the compound from n-type to p-type, thus shifting its Fermi level, or oxidation-reduction properties, and making the generation of free radicals less likely.

Treatments and diagnostic approaches based on the use of nanoparticles are expected to have important benefits for medicine in the future, but the use of nanoparticles also presents significant challenges, particularly regarding impacts on human health. For example, little is known about the fate of nanoparticles that are introduced into the body or whether they have undesirable effects on the body. Extensive clinical trials are needed in order to fully address concerns about the safety and effectiveness of nanoparticles used in medicine. There also are manufacturing problems to be overcome, such as the ability to produce nanoparticles under sterile conditions, which is required for medical applications.

Manufacture of Nanoparticles

Nanoparticles are made by one of three routes: by comminution (the pulverization of materials), such as through industrial milling or natural weathering; by pyrolysis (incineration); or by sol-gel synthesis (the generation of inorganic materials from a colloidal suspension). Comminution is known as a top-down approach, whereas the sol-gel process is a bottom-up approach. Examples of those three processes (comminution, pyrolysis, and sol-gel synthesis) include the production of titania nanoparticles for sunscreens from the minerals anatase and rutile, the production of fullerenes or fumed silica, and the production of synthetic (or Stöber) silica, of other "engineered" oxide nanoparticles, and of quantum dots. For the generation of small nanoparticles, comminution is a very inefficient process.

Detection, Characterization and Isolation

The detection and characterization of nanoparticles present scientists with particular challenges. Being of a size that is at least four to seven times smaller than the wavelength of light means that individual nanoparticles cannot be detected by the human eye, and they are observable under optical microscopes only in liquid samples under certain conditions. Thus, in general, specialized techniques are required to see them, and none of those approaches is currently field-deployable.

Techniques to detect and characterize nanoparticles fall into two categories: direct, or "real space," and indirect, or "reciprocal space." Direct techniques include transmission electron microscopy (TEM), scanning electron microscopy (SEM), and atomic force microscopy (AFM). Those techniques can image nanoparticles, directly measure sizes, and infer shape information, but they are limited to studying only a few particles at a time. There are also significant issues surrounding sample preparation for electron microscopy. In general, however, those techniques can be quite effective for obtaining basic information about a nanoparticle.

Indirect techniques use X-rays or neutron beams and obtain their information by mathematically analyzing the radiation scattered or diffracted by the nanoparticles. The techniques of greatest relevance to nanoscience are small-angle X-ray scattering (SAXS) and small-angle neutron scattering (SANS), along with their surface-specific analogues GISAXS and GISANS, where GI is "grazing incidence", and X-ray or neutron reflectometry (XR/NR). The advantage of those techniques is that they are able to simultaneously sample and average very large numbers of nanoparticles and often do not require any particular sample preparation. Indirect techniques have many applications. For example, in studies of nanoparticles in raw sewage, scientists used SANS measurements, in which neutrons readily penetrated the turbid sewage and scattered strongly from the nanoparticles, to follow the aggregation behaviour of the particles over time.

The isolation of nanoparticles from colloidal and larger matter involves specialized techniques, such as ultra centrifugation and field-flow fractionation. Such laboratory-based techniques are normally coupled to standard spectroscopic instrumentation to enable particular types of chemical characterization.

Nanoparticles in the Environment

Nanoparticles occur naturally in the environment in large volumes. For example, the sea emits an aerosol of salt that ends up floating around in the atmosphere in a range of sizes, from a few nanometres upward, and smoke from volcanoes and fires contains a huge variety of nanoparticles, many of which could be classified as dangerous to human health. Dust from deserts, fields, and so on also has a range of sizes and types of particles, and even trees emit nanoparticles of hydrocarbon compounds such as terpenes (which produce the familiar blue haze seen in forests, from which the Great Smoky Mountains in the United States get their name).

Human-made (anthropogenic) nanoparticles are emitted by large industrial processes, and in modern life it is particles from power stations and from jet aircraft and other vehicles (namely, those powered by internal-combustion engines; car tires are also a factor) that constitute the major fraction of nanoparticle emissions. Types of nanoparticles that are emitted include partially burned hydrocarbons (in soot), ceria (cerium oxide; from vehicle exhaust catalysts), metallic dust (from brake linings), calcium carbonate (in engine lubricating oils), and silica (from car tires). Other sources of nanoparticles to the environment include the semiconductorindustry, domestic and industrial wastewater discharges, the health care industry, and the photographic industry. However, all those emission levels are still considered to be lower than the levels of nanoparticles produced through natural processes. Indeed, recent human-made particles contribute only a small amount to air and water pollution.

Understanding the relationship between nanoparticles and the environment forms an important area of research. There are several mechanisms by which nanoparticles are believed to affect the environment negatively. Two scenarios that are under investigation are the possibilities: (1) that the mobility and sorptive capacity of nanoparticles (natural or human-made) make them potent vectors (carriers) in the transport of chemical pollutants (e.g., phosphorus from sewage and agriculture), particularly in rivers and lakes, and (2) that some nanoparticles are able to reduce the functioning of (and may even disrupt or kill) naturally occurring microbial communities, as well as microbial communities that are employed in industrial processes (e.g., those that are used in sanitation processes, including sewage treatment).

Nanoparticles also can have beneficial impacts on the environment and appear to contribute to natural processes. Thus, in addition to the potential use of nanoparticles to remove chemical contaminants from the environment, scientists are investigating how nanoparticles interact with all life-forms—from fungi to microbes, algae, plants, and higher-order animals. That type of study is essential not only to improving scientists' knowledge of nanoparticles but also to gaining a more complete understanding of life on Earth, since the soil is naturally full of nanoparticles, in a richly diverse environment.

Health Effects of Nanoparticles

Humans have evolved to cope with most naturally occurring nanoparticles. However, some nanoparticles, generated as a result of certain human activities such as tobacco smoking and fires, account for many premature deaths as a result of lung damage. For example, fires from the types of cooking stoves used in developing countries are known to emit fine particles and lead to early mortality, especially among women who routinely work near the stoves.

Laboratory and clinical investigations of the effects of nanoparticles on health have been somewhat controversial and remain largely inconclusive. Most studies in animals have involved nanoparticle inhalation, and the dosages have been very large. The results of those studies have indicated that large quantities of nanoparticles can cause cellular damage in the lungs, with lung cells absorbing the particles and becoming damaged or undergoing genetic mutation. Animal studies involving the ingestion of nanoparticles in food or water suggest that nanoparticles can also affect health in other ways. For example, consumption of the food additive E171, which consists of titanium dioxide nanoparticles, is associated with changes in gut microbiota (bacteria occurring in the gut), potentially contributing to the development of conditions such as inflammatory bowel disease.

In humans, the health effects of typical exposure levels—those that are encountered by most persons during daily activities—remain unknown. Nonetheless, there is a general awareness of the problems that might occur upon excess exposure to nanoparticles, and, thus, most manufacturers of such particles take serious precautions to avoid exposure of their workers. Efforts have been made to educate the public in the use of nanoparticle-containing products. The existence of pressure groups has also helped to ensure nanoparticle safety complianceamong manufacturers. However, nanoparticles offer tremendous potential for new or improved forms of health care treatment. That has spawned a new field of science called nanomedicine.

CARBON NANOTUBE

Carbon nanotubes (CNTs) are tubes made of carbon with diameters typically measured in nanometers. The name carbon nanotubes can refer to tubes with an undetermined carbon wall structure and diameters less than 100 nanometers. Such tubes are thought to have been discovered by Radushkevich and Lukyanovich. The name carbon nanotubes often refers to single-wall carbon nanotubes (SWCNTs) discovered independently by Iijima and Ichihashi and Bethune et al. in carbon arc chambers similar to those used to produce fullerenes. Although not made this way, these tubes can be thought of as cutouts from a two-dimensional hexagonal lattice of carbon atoms rolled up along one of the Bravais lattice vectors of the hexagonal lattice to form a hollow cylinder. This construction, described by Iijima, yields a lattice with helical symmetry of seamlessly bonded carbon atoms on the cylinder surface. Single-wall carbon nanotubes are one of the allotropes of carbon, intermediate between fullerene cages and flat graphene. Besides single-wall carbon nanotubes, carbon nanotubes often refers to multi-wall carbon nanotubes (MWCNTs) consisting of nested single-wall carbon nanotubes that, if not identical, are very similar to Oberlin, Endo and Koyama's long straight and parallel carbon layers cylindrically rolled around a hollow tube. Individual nanotubes naturally align themselves into "ropes" held together by relatively weak van der Waals forces. While nanotubes of other compositions exist, most research has been focused on the carbon ones. Therefore, the "carbon" qualifier is often left implicit in the acronyms, and the names are abbreviated NT, SWNT, and MWNT.

Carbon nanotubes can exhibit remarkable electrical conductivity. They also have exceptional tensile strength and thermal conductivity, because of their nanostructure and strength of the bonds between carbon atoms. In addition, they can be chemically modified. These properties are expected to be valuable in many areas of technology, such as electronics, optics, composite materials (replacing or complementing carbon fibers), nanotechnology, and other applications of materials science. The length of a carbon nanotube produced by common production methods is often not reported, but is much larger than its diameter. Although rare, nanotubes half a meter long have been created, with a length-to-diameter ratio of more than 100,000,000:1. For many purposes, the length of carbon nanotubes can be assumed to be infinite.

The properties of a long and narrow carbon nanotube (for example, whether it is a metal or semiconductor) are largely determined by its diameter and by the "rolling" angle between the main directions of the graphene lattice and the axis of the cylinder. These parameters are constrained so that the type of nanotube can be described by two small integers. Most nanotube types are chiral, meaning that a tube cannot be rotated and translated to match its mirror image. Apart from that, carbon nanotubes are highly symmetrical: every atom in an infinite nanotube is equivalent to any other atom.

The unique strength of carbon nanotubes (or fullerenes in general) is due to orbital hybridization, which causes the bonds between adjacent carbon atoms to be of the sp^2 type. These bonds, which are similar to those of graphene, are stronger than the sp^3 bonds in alkanes and diamond.

Structure of Single-walled Tubes

The structure of an ideal (infinitely long) single-walled carbon nanotube is that of a regular hexagonal lattice drawn on an infinite cylindrical surface, whose vertices are the positions of the carbon atoms. Since the length of the carbon-carbon bonds is fairly fixed, there are constraints on the diameter of the cylinder and the arrangement of the atoms on it.

The Zigzag and Armchair Configurations

In the study of nanotubes, one defines a "zigzag" path on a graphene-like lattice as a path that turns 60 degrees, alternating left and right, after stepping through each bond. It is also conventional to define an "armchair" path as one that makes two left turns of 60 degrees followed by two right turns every four steps.

On some carbon nanotubes, there is a closed zigzag path that goes around the tube. One says that the tube is of the zigzag type or configuration, or simply is a zigzag nanotube. If the tube is instead encircled by a closed armchair path, it is said to be of the armchair type, or an armchair nanotube.

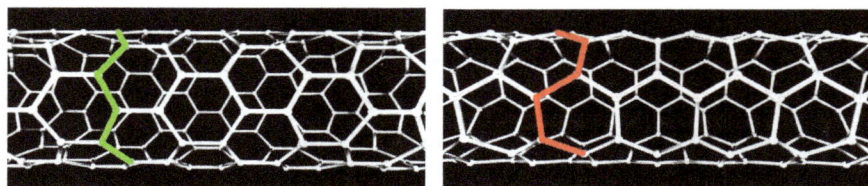

Zigzag nanotube.   Armchair nanotube.

An infinite nanotube that is of the zigzag (or armchair) type consists entirely of closed zigzag (or armchair) paths, connected to each other.

The (n,m) Notation

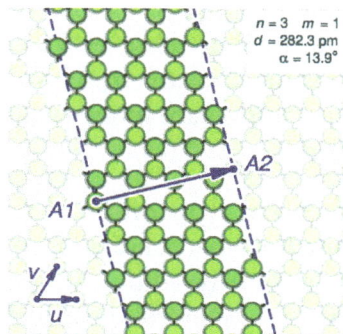

A "sliced and unrolled" representation of a carbon nanotube as a strip of a graphene molecule, overlaid on diagram of the full molecule (faint background). The arrow shows the gap *A2* where the atom *A1* on one edge of the strip would fit in the opposite edge, as the strip is rolled up.

The zigzag and armchair configurations are not the only structures that a single-walled nanotube can have. To describe the structure of a general infinitely long tube, one should imagine it being sliced open by a cut parallel to its axis, that goes through some atom *A*, and then unrolled flat on the plane, so that its atoms and bonds coincide with those of an imaginary graphene sheet—more precisely, with an infinitely long strip of that sheet.

The two halves of the atom *A* will end up on opposite edges of the strip, over two atoms *A1* and *A2* of the graphene. The line from *A1* to *A2* will correspond to the circumference of the cylinder that went through the atom *A*, and will be perpendicular to the edges of the strip.

In the graphene lattice, the atoms can be split into two classes, depending on the directions of their three bonds. Half the atoms have their three bonds directed the same way, and half have their three bonds rotated 180 degrees relative to the first half. The atoms *A1* and *A2*, which correspond to the same atom *A* on the cylinder, must be in the same class.

It follows that the circumference of the tube and the angle of the strip are not arbitrary, because they are constrained to the lengths and directions of the lines that connect pairs of graphene atoms in the same class.

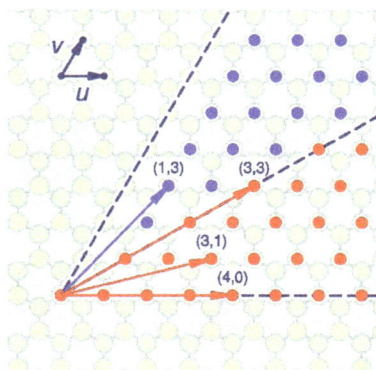

The basis vectors *u* and *v* of the relevant sub-lattice, the (n,m) pairs that define non-isomorphic carbon nanotube structures (red dots), and the pairs that define the enantiomers of the chiral ones (blue dots).

Let u and v be two linearly independent vectors that connect the graphene atom $A1$ to two of its nearest atoms with the same bond directions. That is, if one numbers consecutive carbons around a graphene cell with C1 to C6, then u can be the vector from C1 to C3, and v be the vector from C1 to C5. Then, for any other atom $A2$ with same class as $A1$, the vector from $A1$ to $A2$ can be written as a linear combination $n\,u + m\,v$, where n and m are integers. And, conversely, each pair of integers (n,m) defines a possible position for $A2$.

Given n and m, one can reverse this theoretical operation by drawing the vector w on the graphene lattice, cutting a strip of the latter along lines perpendicular to w through its endpoints $A1$ and $A2$, and rolling the strip into a cylinder so as to bring those two points together. If this construction is applied to a pair $(k,0)$, the result is a zigzag nanotube, with closed zigzag paths of $2k$ atoms. If it is applied to a pair (k,k), one obtains an armchair tube, with closed armchair paths of $4k$ atoms.

Nanotube Types

Moreover, the structure of the nanotube is not changed if the strip is rotated by 60 degrees clockwise around $A1$ before applying the hypothetical reconstruction above. Such a rotation changes the corresponding pair (n,m) to the pair $(-2m,n+m)$.

It follows that many possible positions of $A2$ relative to $A1$ — that is, many pairs (n,m) — correspond to the same arrangement of atoms on the nanotube. That is the case, for example, of the six pairs $(1,2)$, $(-2,3)$, $(-3,1)$, $(-1,-2)$, $(2,-3)$, and $(3,-1)$. In particular, the pairs $(k,0)$ and $(0,k)$ describe the same nanotube geometry.

These redundancies can be avoided by considering only pairs (n,m) such that $n > 0$ and $m \geq 0$; that is, where the direction of the vector w lies between those of u (inclusive) and v (exclusive). It can be verified that every nanotube has exactly one pair (n,m) that satisfies those conditions, which is called the tube's type. Conversely, for every type there is a hypothetical nanotube. In fact, two nanotubes have the same type if and only if one can be conceptually rotated and translated so as to match the other exactly.

Instead of the type (n,m), the structure of a carbon nanotube can be specified by giving the length of the vector w (that is, the circumference of the nanotube), and the angle α between the directions of u and w, which may range from 0 (inclusive) to 60 degrees clockwise (exclusive). If the diagram is drawn with u horizontal, the latter is the tilt of the strip away from the vertical.

Here are some unrolled nanotube diagrams:

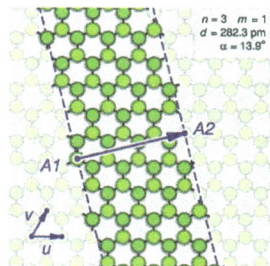

Chiral nanotube of the (3,1) type.

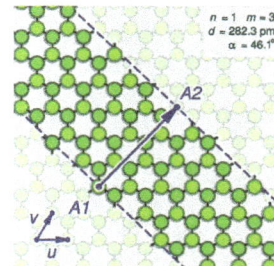

Chiral nanotube of the (1,3) type, mirror image of the (3,1) type.

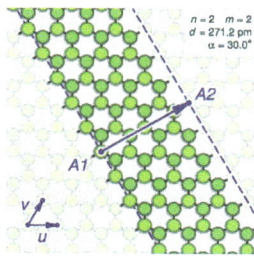

Nanotube of the (2,2) type, the narrowest "armchair" one.

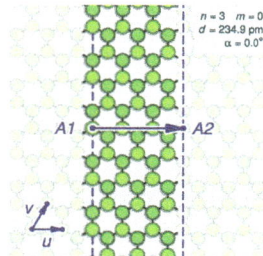

Nanotube of the (3,0) type, the narrowest "zigzag" one.

Chiralty and Mirror Symmetry

A nanotube is chiral if it has type (n,m), with m > 0 and m ≠ n; then its enantiomer (mirror image) has type (m,n), which is different from (n,m). This operation corresponds to mirroring the unrolled strip about the line L through A1 that makes an angle of 30 degrees clockwise from the direction of the u vector (that is, with the direction of the vector u+v). The only types of nanotubes that are achiral are the (k,0) "zigzag" tubes and the (k,k) "armchair" tubes.

If two enantiomers are to be considered the same structure, then one may consider only types (n,m) with $0 \le m \le n$ and n > 0. Then the angle α between u and w, which may range from 0 to 30 degrees (inclusive both), is called the "chiral angle" of the nanotube.

Circumference and Diameter

From n and m one can also compute the circumference c, which is the length of the vector w, which turns out to be:

$$c = |u|\sqrt{(n^2 + nm + m^2)} \approx 246\sqrt{((n+m)^2 - nm)}$$

in picometres. The diameter d of the tube is then c/π, that is:

$$d \approx 78.3\sqrt{((n+m)^2 - nm)}$$

also in picometres. (These formulas are only approximate, especially for small n and m where the bonds are strained; and they do not take into account the thickness of the wall.)

The tilt angle α between u and w and the circumference c are related to the type indices n and m by:

$$\alpha = \arg(n+m/2, m\sqrt{3}/2) = \arccos\frac{n+m/2}{c}$$

where arg (x,y) is the clockwise angle between the X-axis and the vector (x,y); a function that is available in many programming languages as atan2(y,x). Conversely, given c and α, one can get the type (n,m) by the formulas:

$$m = \frac{2c}{\sqrt{3}}\sin\alpha \qquad n = c\cos\alpha - \frac{m}{2}$$

which must evaluate to integers.

Physical Limits

Narrowest Nanotubes

If n and m are too small, the structure described by the pair (n,m) will describe a molecule that cannot be reasonably called a "tube", and may not even be stable. For example, the structure theoretically described by the pair (1,0) (the limiting "zigzag" type) would be just a chain of carbons. That is a real molecule, the carbyne; which has some characteristics of nanotubes (such as orbital hybridization, high tensile strength, etc.) — but has no hollow space, and may not be obtainable as a condensed phase. The pair (2,0) would theoretically yield a chain of fused 4-cycles; and (1,1), the limiting "armchair" structure, would yield a chain of bi-connected 4-rings. These structures may not be realizable.

The thinnest carbon nanotube proper is the armchair structure with type (2,2), which has a diameter of 0.3 nm. This nanotube was grown inside a multi-walled carbon nanotube. Assigning of the carbon nanotube type was done by a combination of high-resolution transmission electron microscopy (HRTEM), Raman spectroscopy, and density functional theory (DFT) calculations.

The thinnest *freestanding* single-walled carbon nanotube is about 0.43 nm in diameter. Researchers suggested that it can be either (5,1) or (4,2) SWCNT, but the exact type of the carbon nanotube remains questionable. (3,3), (4,3), and (5,1) carbon nanotubes (all about 0.4 nm in diameter) were unambiguously identified using aberration-corrected high-resolution transmission electron microscopy inside double-walled CNTs.

Here are some tube types that are "degenerate" for being too narrow:

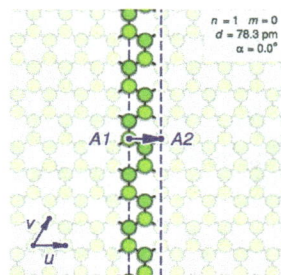

Denegerate "zigzag" tube type (1,0).   Degenerate "zigzag" tube type (2,0).

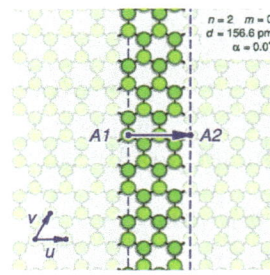

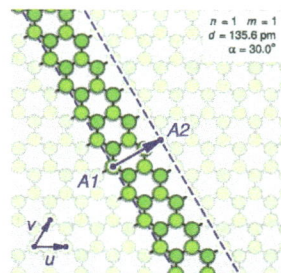

Degenerate "armchair" tube type (1,1).   Possibly degenerate chiral tube type (2,1).

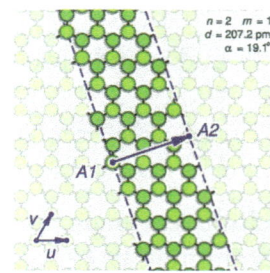

Length

The observation of the *longest* carbon nanotubes grown so far, around 1/2 meter (550 mm long), was reported in 2013. These nanotubes were grown on silicon substrates using an improved chemical vapor deposition (CVD) method and represent electrically uniform arrays of single-walled carbon nanotubes.

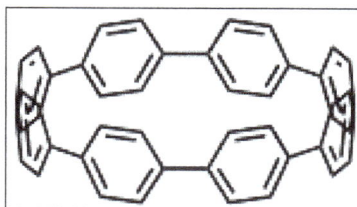

Cycloparaphenylene.

The *shortest* carbon nanotube can be considered to be the organic compound cycloparaphenylene, which was synthesized in 2008.

Density

The *highest density* of CNTs was achieved in 2013, grown on a conductive titanium-coated copper surface that was coated with co-catalysts cobalt and molybdenum at lower than typical temperatures of 450 °C. The tubes averaged a height of 380 nm and a mass density of 1.6 g cm^{-3}. The material showed ohmic conductivity (lowest resistance ~22 kΩ).

Variants

There is no consensus on some terms describing carbon nanotubes in scientific literature: both "-wall" and "-walled" are being used in combination with "single", "double", "triple", or "multi", and the letter C is often omitted in the abbreviation, for example, multi-walled carbon nanotube (MWNT). International Standards Organization uses single-wall or multi-wall in its documents.

Multi-walled Nanotube

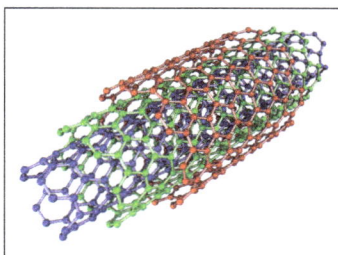

Triple-walled armchair carbon nanotube.

Multi-walled nanotubes (MWNTs) consist of multiple rolled layers (concentric tubes) of graphene. There are two models that can be used to describe the structures of multi-walled nanotubes. In the *Russian Doll* model, sheets of graphite are arranged in concentric cylinders, e.g., a (0,8) single-walled nanotube (SWNT) within a larger (0,17) single-walled nanotube. In the *Parchment* model, a single sheet of graphite is rolled in around itself, resembling a scroll of parchment or a rolled newspaper. The interlayer distance in multi-walled nanotubes is close to the distance between graphene layers in graphite, approximately 3.4 Å. The Russian Doll structure is observed more commonly. Its individual shells can be described as SWNTs, which can be metallic or semiconducting. Because of statistical probability and restrictions on the relative diameters of the individual tubes, one of the shells, and thus the whole MWNT, is usually a zero-gap metal.

Double-walled carbon nanotubes (DWNTs) form a special class of nanotubes because their morphology and properties are similar to those of SWNTs but they are more resistant to chemicals. This is especially important when it is necessary to graft chemical functions to the surface of the nanotubes (functionalization) to add properties to the CNT. Covalent functionalization of SWNTs will break some C=C double bonds, leaving "holes" in the structure on the nanotube and thus modifying both its mechanical and electrical properties. In the case of DWNTs, only the outer wall is modified. DWNT synthesis on the gram-scale by the CCVD technique was first proposed in 2003 from the selective reduction of oxide solutions in methane and hydrogen.

The telescopic motion ability of inner shells and their unique mechanical properties will permit the use of multi-walled nanotubes as the main movable arms in upcoming nanomechanical devices. The retraction force that occurs to telescopic motion is caused by the Lennard-Jones interaction between shells, and its value is about 1.5 nN.

Junctions and Crosslinking

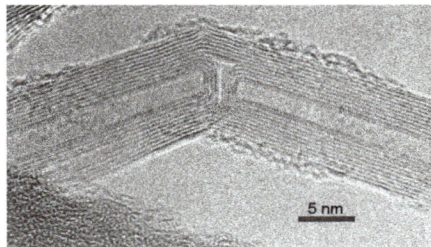

Transmission electron microscope image of carbon nanotube junction.

Junctions between two or more nanotubes have been widely discussed theoretically. Such junctions are quite frequently observed in samples prepared by arc discharge as well as by chemical vapor deposition. The electronic properties of such junctions were first considered theoretically by Lambin et al., who pointed out that a connection between a metallic tube and a semiconducting one would represent a nanoscale heterojunction. Such a junction could therefore form a component of a nanotube-based electronic circuit. The adjacent image shows a junction between two multiwalled nanotubes.

Junctions between nanotubes and graphene have been considered theoretically and studied experimentally. Nanotube-graphene junctions form the basis of pillared graphene, in which parallel graphene sheets are separated by short nanotubes. Pillared graphene represents a class of three-dimensional carbon nanotube architectures.

3D carbon scaffolds.

Recently, several studies have highlighted the prospect of using carbon nanotubes as building blocks to fabricate three-dimensional macroscopic (>100 nm in all three dimensions) all-carbon devices. Lalwani et al. have reported a novel radical-initiated thermal crosslinking method to fabricate macroscopic, free-standing, porous, all-carbon scaffolds using single- and multi-walled carbon nanotubes as building blocks. These scaffolds possess macro-, micro-, and nano-structured pores, and the porosity can be tailored for specific applications. These 3D all-carbon scaffolds/architectures may be used for the fabrication of the next generation of energy storage, supercapacitors, field emission transistors, high-performance catalysis, photovoltaics, and biomedical devices, implants, and sensors.

Other Morphologies

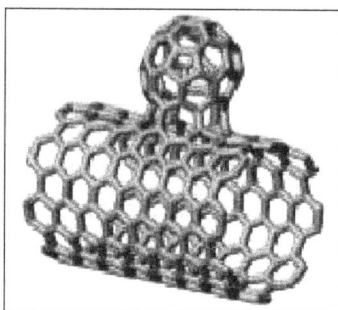

A stable nanobud structure.

Carbon nanobuds are a newly created material combining two previously discovered allotropes of carbon: carbon nanotubes and fullerenes. In this new material, fullerene-like "buds" are covalently bonded to the outer sidewalls of the underlying carbon nanotube. This hybrid material has useful properties of both fullerenes and carbon nanotubes. In particular, they have been found to be exceptionally good field emitters. In composite materials, the attached fullerene molecules may function as molecular anchors preventing slipping of the nanotubes, thus improving the composite's mechanical properties.

A carbon peapod is a novel hybrid carbon material which traps fullerene inside a carbon nanotube. It can possess interesting magnetic properties with heating and irradiation. It can also be applied as an oscillator during theoretical investigations and predictions.

In theory, a nanotorus is a carbon nanotube bent into a torus (doughnut shape). Nanotori are predicted to have many unique properties, such as magnetic moments 1000 times larger than that previously expected for certain specific radii. Properties such as magnetic moment, thermal stability, etc. vary widely depending on the radius of the torus and the radius of the tube.

Graphenated carbon nanotubes are a relatively new hybrid that combines graphitic foliates grown along the sidewalls of multiwalled or bamboo style CNTs. The foliate density can vary as a function of deposition conditions (e.g., temperature and time) with their structure ranging from a few layers of graphene (< 10) to thicker, more graphite-like. The fundamental advantage of an integrated graphene-CNT structure is the high surface area three-dimensional framework of the CNTs coupled with the high edge density of graphene. Depositing a high density of graphene foliates along the length of aligned CNTs can significantly increase the total charge capacity per unit of nominal area as compared to other carbon nanostructures.

Cup-stacked carbon nanotubes (CSCNTs) differ from other quasi-1D carbon structures, which normally behave as quasi-metallic conductors of electrons. CSCNTs exhibit semiconducting behavior because of the stacking microstructure of graphene layers.

Properties

Many properties of single-walled carbon nanotubes depend significantly on the (n,m) type, and this dependence is non-monotonic. In particular, the band gap can vary from zero to about 2 eV and the electrical conductivity can show metallic or semiconducting behavior.

Mechanical

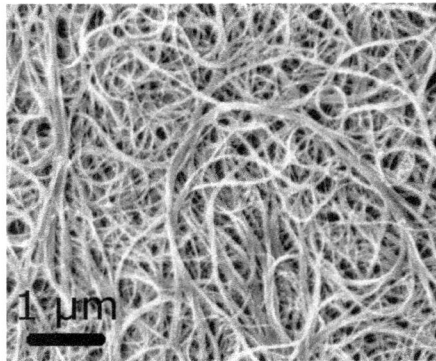

A scanning electron microscopy image of carbon nanotubes bundles.

Carbon nanotubes are the strongest and stiffest materials yet discovered in terms of tensile strength and elastic modulus. This strength results from the covalent sp^2 bonds formed between the individual carbon atoms. In 2000, a multiwalled carbon nanotube was tested to have a tensile strength of 63 gigapascals (9,100,000 psi). (For illustration, this translates into the ability to endure tension of a weight equivalent to 6,422 kilograms-force (62,980 N; 14,160 lbf) on a cable with cross-section of 1 square millimetre (0.0016 sq in)). Further studies, such as one conducted in 2008, revealed that individual CNT shells have strengths of up to ≈100 gigapascals (15,000,000 psi), which is in agreement with quantum/atomistic models. Because carbon nanotubes have a low density for a solid of 1.3 to 1.4 g/cm^3, its specific strength of up to 48,000 kN·m·kg^{-1} is the best of known materials, compared to high-carbon steel's 154 kN·m·kg^{-1}.

Although the strength of individual CNT shells is extremely high, weak shear interactions between adjacent shells and tubes lead to significant reduction in the effective strength of multiwalled carbon nanotubes and carbon nanotube bundles down to only a few GPa. This limitation has been recently addressed by applying high-energy electron irradiation, which crosslinks inner shells and tubes, and effectively increases the strength of these materials to ≈60 GPa for multiwalled carbon nanotubes and ≈17 GPa for double-walled carbon nanotube bundles. CNTs are not nearly as strong under compression. Because of their hollow structure and high aspect ratio, they tend to undergo buckling when placed under compressive, torsional, or bending stress.

On the other hand, there was evidence that in the radial direction they are rather soft. The first transmission electron microscope observation of radial elasticity suggested that even van der Waals forces can deform two adjacent nanotubes. Later, nanoindentations with an atomic force microscope were performed by several groups to quantitatively measure radial elasticity of multi-walled carbon nanotubes and tapping/contact mode atomic force microscopy was also performed on single-walled carbon nanotubes. Young's modulus of on the order of several GPa showed that CNTs are in fact very soft in the radial direction.

Electrical

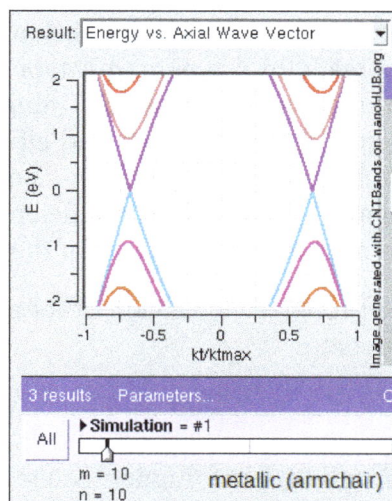

Band structures computed using tight binding approximation for (6,0) CNT (zigzag, metallic), (10,2) CNT (semiconducting) and (10,10) CNT (armchair, metallic).

Unlike graphene, which is a two-dimensional semimetal, carbon nanotubes are either metallic or semiconducting along the tubular axis. For a given (n,m) nanotube, if $n = m$, the nanotube is metallic; if $n - m$ is a multiple of 3 and n ≠ m and nm ≠ 0, then the nanotube is quasi-metallic with a very small band gap, otherwise the nanotube is a moderate semiconductor. Thus, all armchair ($n = m$) nanotubes are metallic, and nanotubes (6,4), (9,1), etc. are semiconducting. Carbon nanotubes are not semimetallic because the degenerate point (the point where the π [bonding] band meets the π^* [anti-bonding] band, at which the energy goes to zero) is slightly shifted away from the K point in the Brillouin zone because of the curvature of the tube surface, causing hybridization between the σ^* and π^* anti-bonding bands, modifying the band dispersion.

The rule regarding metallic versus semiconductor behavior has exceptions because curvature effects in small-diameter tubes can strongly influence electrical properties. Thus, a (5,0) SWCNT that should be semiconducting in fact is metallic according to the calculations. Likewise, zigzag and chiral SWCNTs with small diameters that should be metallic have a finite gap (armchair nanotubes remain metallic). In theory, metallic nanotubes can carry an electric current density of 4×10^9 A/cm², which is more than 1,000 times greater than those of metals such as copper, where for copper interconnects, current densities are limited by electromigration. Carbon nanotubes are thus being explored as interconnects and conductivity-enhancing components in composite materials, and many groups are attempting to commercialize highly conducting electrical wire assembled from individual carbon nanotubes. There are significant challenges to be overcome however, such as undesired current saturation under voltage, and the much more resistive nanotube-to-nanotube junctions and impurities, all of which lower the electrical conductivity of the macroscopic nanotube wires by orders of magnitude, as compared to the conductivity of the individual nanotubes.

Because of its nanoscale cross-section, electrons propagate only along the tube's axis. As a result, carbon nanotubes are frequently referred to as one-dimensional conductors. The maximum electrical conductance of a single-walled carbon nanotube is $2G_0$, where $G_0 = 2e^2/h$ is the conductance of a single ballistic quantum channel.

Because of the role of the π-electron system in determining the electronic properties of graphene, doping in carbon nanotubes differs from that of bulk crystalline semiconductors from the same group of the periodic table (e.g., silicon). Graphitic substitution of carbon atoms in the nanotube wall by boron or nitrogen dopants leads to p-type and n-type behavior, respectively, as would be expected in silicon. However, some non-substitutional (intercalated or adsorbed) dopants introduced into a carbon nanotube, such as alkali metals and electron-rich metallocenes, result in n-type conduction because they donate electrons to the π-electron system of the nanotube. By contrast, π-electron acceptors such as $FeCl_3$ or electron-deficient metallocenes function as p-type dopants because they draw π-electrons away from the top of the valence band.

Intrinsic superconductivity has been reported, although other experiments found no evidence of this, leaving the claim a subject of debate.

Optical

Carbon nanotubes have useful absorption, photoluminescence (fluorescence), and Raman spectroscopy properties. Spectroscopic methods offer the possibility of quick and non-destructive characterization of relatively large amounts of carbon nanotubes. There is a strong demand for such characterization from the industrial point of view: numerous parameters of nanotube synthesis can be changed, intentionally or unintentionally, to alter the nanotube quality. As shown below, optical absorption, photoluminescence, and Raman spectroscopies allow quick and reliable characterization of this "nanotube quality" in terms of non-tubular carbon content, structure (chirality) of the produced nanotubes, and structural defects. These features determine nearly any other properties such as optical, mechanical, and electrical properties.

Carbon nanotubes are unique "one-dimensional systems" which can be envisioned as rolled single sheets of graphite (or more precisely graphene). This rolling can be done at different angles and curvatures resulting in different nanotube properties. The diameter typically varies in the range 0.4–40 nm (i.e., "only" ~100 times), but the length can vary ~100,000,000,000 times, from 0.14 nm to 55.5 cm. The nanotube aspect ratio, or the length-to-diameter ratio, can be as high as 132,000,000:1, which is unequalled by any other material. Consequently, all the properties of the carbon nanotubes relative to those of typical semiconductors are extremely anisotropic (directionally dependent) and tunable.

Whereas mechanical, electrical, and electrochemical (supercapacitor) properties of the carbon nanotubes are well established and have immediate applications, the practical use of optical properties is yet unclear. The aforementioned tunability of properties is potentially useful in optics and photonics. In particular, light-emitting diodes (LEDs) and photo-detectors based on a single nanotube have been produced in the lab. Their unique feature is not the efficiency, which is yet relatively low, but the narrow selectivity in the wavelength of emission and detection of light and the possibility of its fine tuning through the nanotube structure. In addition, bolometer and optoelectronic memory devices have been realised on ensembles of single-walled carbon nanotubes.

Crystallographic defects also affect the tube's electrical properties. A common result is lowered conductivity through the defective region of the tube. A defect in armchair-type tubes (which can conduct electricity) can cause the surrounding region to become semiconducting, and single monatomic vacancies induce magnetic properties.

Thermal

All nanotubes are expected to be very good thermal conductors along the tube, exhibiting a property known as "ballistic conduction", but good insulators lateral to the tube axis. Measurements show that an individual SWNT has a room-temperature thermal conductivity along its axis of about 3500 W·m^{-1}·K^{-1}; compare this to copper, a metal well known for its good thermal conductivity, which transmits 385 W·m^{-1}·K^{-1}. An individual SWNT has a room-temperature thermal conductivity across its axis (in the radial direction) of about 1.52 W·m^{-1}·K^{-1}, which is about as thermally conductive as soil. Macroscopic assemblies of nanotubes such as films or fibres have reached up to 1500 W·m^{-1}·K^{-1} so far. Networks composed of nanotubes demonstrate different values of thermal conductivity, from the level of thermal insulation with the thermal conductivity of 0.1 W·m^{-1}·K^{-1} to such high values. That is dependent on the amount of contribution to the thermal resistance of the system caused by the presence of impurities, misalignments and other factors. The temperature stability of carbon nanotubes is estimated to be up to 2800 °C in vacuum and about 750 °C in air.

Crystallographic defects strongly affect the tube's thermal properties. Such defects lead to phonon scattering, which in turn increases the relaxation rate of the phonons. This reduces the mean free path and reduces the thermal conductivity of nanotube structures. Phonon transport simulations indicate that substitutional defects such as nitrogen or boron will primarily lead to scattering of high-frequency optical phonons. However, larger-scale defects such as Stone Wales defects cause phonon scattering over a wide range of frequencies, leading to a greater reduction in thermal conductivity.

Synthesis

Techniques have been developed to produce nanotubes in sizable quantities, including arc discharge, laser ablation, chemical vapor deposition (CVD) and high-pressure carbon monoxide disproportionation (HiPCO). Among these arc discharge, laser ablation, chemical vapor deposition (CVD) are batch by batch process and HiPCO is gas phase continuous process. Most of these processes take place in a vacuum or with process gases. The CVD growth method is popular, as it yields high quantity and has a degree of control over diameter, length and morphology. Using particulate catalysts, large quantities of nanotubes can be synthesized by these methods, but achieving the repeatability becomes a major problem with CVD growth. The HiPCO process advances in catalysis and continuous growth are making CNTs more commercially viable. The HiPCO process helps in producing high purity single walled carbon nanotubes in higher quantity. The HiPCO reactor operates at high temperature 900-1100 °C and high pressure ~30-50 bar. It uses carbon monoxide as the carbon source and Nickel/ iron penta carbonyl as catalyst. These catalyst acts as the nucleation site for the nanotubes to grow.

Vertically aligned carbon nanotube arrays are also grown by thermal chemical vapor deposition. A substrate (quartz, silicon, stainless steel, etc.) is coated with a catalytic metal (Fe, Co, Ni) layer. Typically that layer is iron, and is deposited via sputtering to a thickness of 1–5 nm. A 10–50 nm underlayer of alumina is often also put down on the substrate first. This imparts controllable wetting and good interfacial properties. When the substrate is heated to the growth temperature (~700 °C), the continuous iron film breaks up into small islands each island then nucleates a carbon nanotube. The sputtered thickness controls the island size, and this in turn determines the nanotube diameter. Thinner iron layers drive down the diameter of the islands, and they drive down the diameter of the nanotubes grown. The amount of time that the metal island can sit at the growth temperature is limited, as they are mobile, and can merge into larger (but fewer) islands. Annealing at the growth temperature reduces the site density (number of CNT/mm^2) while increasing the catalyst diameter.

The as-prepared carbon nanotubes always have impurities such as other forms of carbon (amorphous carbon, fullerene, etc.) and non-carbonaceous impurities (metal used for catalyst). These impurities need to be removed to make use of the carbon nanotubes in applications.

Modeling

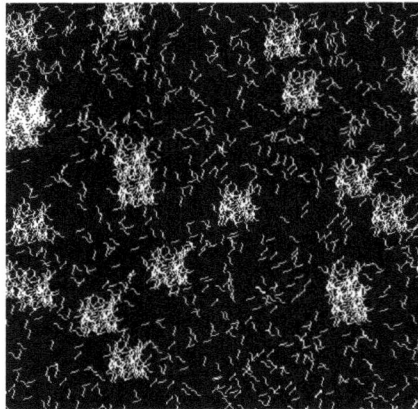

Computer simulated microstructures with agglomeration regions.

Carbon nanotubes are modelled in a similar manner as traditional composites in which a reinforcement phase is surrounded by a matrix phase. Ideal models such as cylinderical, hexagonal and square models are common. The size of the micromechanics model is highly function of the studied mechanical properties. The concept of representative volume element (RVE) is used to determine the appropriate size and configuration of computer model to replicate the actual behavior of CNT reinforced nanocomposite. Depending on the material property of interest (thermal, electrical, modulus, creep), one RVE might predict the property better than the alternatives. While the implementation of ideal model is comutationally efficient, they do not represent microstructural features observed in scanning electron microscopy of actual nanocomposites. To incorporate realistic modeling, computer models are also generated to incorporate variability such as waviness, orientation and agglomeration of multiwall or single wall carbon nanotubes.

Metrology

There are many metrology standards and reference materials available for carbon nanotubes.

For single-wall carbon nanotubes, ISO/TS 10868 describes a measurement method for the diameter, purity, and fraction of metallic nanotubes through optical absorption spectroscopy, while ISO/TS

10797 and ISO/TS 10798 establish methods to characterize the morphology and elemental composition of single-wall carbon nanotubes, using transmission electron microscopy and scanning electron microscopy respectively, coupled with energy dispersive X-ray spectrometry analysis.

NIST SRM 2483 is a soot of single-wall carbon nanotubes used as a reference material for elemental analysis, and was characterized using thermogravimetric analysis, prompt gamma activation analysis, induced neutron activation analysis, inductively coupled plasma mass spectroscopy, resonant Raman scattering, UV-visible-near infrared fluorescence spectroscopy and absorption spectroscopy, scanning electron microscopy, and transmission electron microscopy. The Canadian National Research Council also offers a certified reference material SWCNT-1 for elemental analysis using neutron activation analysis and inductively coupled plasma mass spectroscopy. NIST RM 8281 is a mixture of three lengths of single-wall carbon nanotube.

For multiwall carbon nanotubes, ISO/TR 10929 identifies the basic properties and the content of impurities, while ISO/TS 11888 describes morphology using scanning electron microscopy, transmission electron microscopy, viscometry, and light scattering analysis. ISO/TS 10798 is also valid for multiwall carbon nanotubes.

Chemical Modification

Carbon nanotubes can be functionalized to attain desired properties that can be used in a wide variety of applications. The two main methods of carbon nanotube functionalization are covalent and non-covalent modifications. Because of their apparent hydrophobic nature, carbon nanotubes tend to agglomerate hindering their dispersion in solvents or viscous polymer melts. The resulting nanotube bundles or aggregates reduce the mechanical performance of the final composite. The surface of the carbon nanotubes can be modified to reduce the hydrophobicity and improve interfacial adhesion to a bulk polymer through chemical attachment.

Also surface of carbon nanotubes can be fluorinated or halofluorinated by CVD-method with fluorocarbons, hydro- or halofluorocarbons by heating while in contact of such carbon material with fluoroorganic substance to form partially fluorinated carbons (so called Fluocar materials) with grafted (halo) fluoroalkyl functionality.

Applications

A primary obstacle for applications of carbon nanotubes has been their cost. Prices for single-walled nanotubes declined from around $1500 per gram as of 2000 to retail prices of around $50 per gram of as-produced 40–60% by weight SWNTs as of March 2010. As of 2016, the retail price of as-produced 75% by weight SWNTs was $2 per gram. SWNTs are forecast to make a large impact in electronics applications by 2020 according to *The Global Market for Carbon Nanotubes* report.

Current

Current use and application of nanotubes has mostly been limited to the use of bulk nanotubes, which is a mass of rather unorganized fragments of nanotubes. Bulk nanotube materials may never achieve a tensile strength similar to that of individual tubes, but such composites may, nevertheless, yield strengths sufficient for many applications. Bulk carbon nanotubes have already been used as composite fibers in polymers to improve the mechanical, thermal and electrical properties of the bulk product.

- Easton-Bell Sports, Inc. have been in partnership with Zyvex Performance Materials, using CNT technology in a number of their bicycle components – including flat and riser handle-bars, cranks, forks, seatposts, stems and aero bars.

- Amroy Europe Oy manufactures Hybtonite carbon nanoepoxy resins where carbon nano-tubes have been chemically activated to bond to epoxy, resulting in a composite material that is 20% to 30% stronger than other composite materials. It has been used for wind turbines, marine paints and a variety of sports gear such as skis, ice hockey sticks, baseball bats, hunting arrows, and surfboards.

- Surrey NanoSystems synthesises carbon nanotubes to create vantablack.

Other current applications include:

- Tips for atomic force microscope probes.

- In tissue engineering, carbon nanotubes can act as scaffolding for bone growth.

Under Development

Current research for modern applications include:

- Using carbon nanotubes as a scaffold for diverse microfabrication techniques.

- Energy dissipation in self-organized nanostructures under influence of an electric field.

- Using carbon nanotubes for environmental monitoring due to their active surface area and their ability to absorb gases.

- Jack Andraka used carbon nanotubes in his pancreatic cancer test. His method of testing won the Intel International Science and Engineering Fair Gordon E. Moore Award in the spring of 2012.

- The Boeing Company has patented the use of carbon nanotubes for structural health moni-toring of composites used in aircraft structures. This technology will greatly reduce the risk of an in-flight failure caused by structural degradation of aircraft.

- Zyvex Technologies has also built a 54' maritime vessel, the Piranha Unmanned Surface Vessel, as a technology demonstrator for what is possible using CNT technology. CNTs help improve the structural performance of the vessel, resulting in a lightweight 8,000 lb boat that can carry a payload of 15,000 lb over a range of 2,500 miles.

Carbon nanotubes can serve as additives to various structural materials. For instance, nanotubes form a tiny portion of the material(s) in some (primarily carbon fiber) baseball bats, golf clubs, car parts, or damascus steel.

Potential

The strength and flexibility of carbon nanotubes makes them of potential use in controlling other nanoscale structures, which suggests they will have an important role in nanotechnology engineering. The highest tensile strength of an individual multi-walled carbon nanotube has been tested to be 63 GPa. Carbon nanotubes were found in Damascus steel from the 17th century, possibly helping to account for the legendary strength of the swords made of it. Recently, several studies have highlighted the prospect of using carbon nanotubes as building blocks to fabricate three-dimensional macroscopic (>1mm in all three dimensions) all-carbon devices. Lalwani et al. have reported a novel radical initiated thermal crosslinking method to fabricated macroscopic, free-standing, porous, all-carbon scaffolds using single- and multi-walled carbon nanotubes as building blocks. These scaffolds possess macro-, micro-, and nano- structured pores and the porosity can be tailored for specific applications. These 3D all-carbon scaffolds/architectures may be used for the fabrication of the next generation of energy storage, supercapacitors, field emission transistors, high-performance catalysis, photovoltaics, and biomedical devices and implants.

CNTs are potential candidates for future via and wire material in nano-scale VLSI circuits. Eliminating electromigration reliability concerns that plague today's Cu interconnects, isolated (single and multi-wall) CNTs can carry current densities in excess of 1000 MA/sq-cm without electromigration damage.

Single-walled nanotubes are likely candidates for miniaturizing electronics. The most basic building block of these systems is an electric wire, and SWNTs with diameters of an order of a nanometer can be excellent conductors. One useful application of SWNTs is in the development of the first intermolecular field-effect transistors (FET). The first intermolecular logic gate using SWCNT FETs was made in 2001. A logic gate requires both a p-FET and an n-FET. Because SWNTs are p-FETs when exposed to oxygen and n-FETs otherwise, it is possible to expose half of an SWNT to oxygen and protect the other half from it. The resulting SWNT acts as a *not* logic gate with both p- and n-type FETs in the same molecule. Large quantities of pure CNTs can be made into a free-standing sheet or film by surface-engineered tape-casting (SETC) fabrication technique which is a scalable method to fabricate flexible and foldable sheets with superior properties. Another reported form factor is CNT fiber (a.k.a. filament) by wet spinning. The fiber is either directly spun from the synthesis pot or spun from pre-made dissolved CNTs. Individual fibers can be turned into a yarn. Apart from its strength and flexibility, the main advantage is making an electrically conducting yarn. The electronic properties of individual CNT fibers (i.e. bundle of individual CNT) are governed by the two-dimensional structure of CNTs. The fibers were measured to have a resistivity only one order of magnitude higher than metallic conductors at 300K. By further optimizing the CNTs and CNT fibers, CNT fibers with improved electrical properties could be developed.

CNT-based yarns are suitable for applications in energy and electrochemical water treatment when coated with an ion-exchange membrane. Also, CNT-based yarns could replace copper as a winding material. Pyrhönen et al. have built a motor using CNT winding.

GRAPHENE

Graphene is an allotrope of carbon in the form of a single layer of atoms in a two-dimensional hexagonal lattice in which one atom forms each vertex. It is the basic structural element of other allotropes, including graphite, charcoal, carbon nanotubes and fullerenes. It can also be considered as an indefinitely large aromatic molecule, the ultimate case of the family of flat polycyclic aromatic hydrocarbons.

Graphene has a unique set of properties which set it apart from other materials. In proportion to its thickness, it is about 100 times stronger than the strongest steel. It conducts heat and electricity very efficiently and is nearly transparent. Graphene also shows a large and nonlinear diamagnetism, even greater than graphite, and can be levitated by Nd-Fe-B magnets. Researchers have identified the bipolar transistor effect, ballistic transport of charges and large quantum oscillations in the material.

Scientists have theorized about graphene for decades. It has likely been unknowingly produced in small quantities for centuries, through the use of pencils and other similar applications of graphite. It was originally observed in electron microscopes in 1962, but only studied while supported on metal surfaces. The material was later rediscovered, isolated and characterized in 2004 by Andre Geim and Konstantin Novoselov at the University of Manchester. Research was informed by existing theoretical descriptions of its composition, structure and properties. High-quality graphene proved to be surprisingly easy to isolate, making more research possible. This work resulted in the two winning the Nobel Prize in Physics in 2010 "for groundbreaking experiments regarding the two-dimensional material graphene".

The global market for graphene is reported to have reached $9 million by 2012, with most of the demand from research and development in semiconductor, electronics, battery energy and composites.

"Graphene" is a combination of "graphite" and the suffix -ene, named by Hanns-Peter Boehm, who described single-layer carbon foils in 1962.

The term *graphene* first appeared in 1987 to describe single sheets of graphite as a constituent of graphite intercalation compounds (GICs); conceptually a GIC is a crystalline salt of the intercalant and graphene. The term was also used in early descriptions of carbon nanotubes, as well as for epitaxial graphene and polycyclic aromatic hydrocarbons. Graphene can be considered an "infinite alternant" (only six-member carbon ring) polycyclic aromatic hydrocarbon (PAH).

The IUPAC compendium of technology states: "previously, descriptions such as graphite layers, carbon layers, or carbon sheets have been used for the term graphene... it is incorrect to use for a single layer a term which includes the term graphite, which would imply a three-dimensional structure. The term graphene should be used only when the reactions, structural relations or other properties of individual layers are discussed."

Geim defined "isolated or free-standing graphene" as "graphene is a single atomic plane of graphite, which – and this is essential – is sufficiently isolated from its environment to be considered free-standing." This definition is narrower than the IUPAC definition and refers to cloven, transferred, and suspended graphene. Other forms of graphene, such as graphene grown on various metals, can become free-standing if, for example, suspended or transferred to silicon dioxide (SiO_2) or silicon carbide.

Properties

Graphene has a theoretical specific surface area (SSA) of 2630 m²/g. This is much larger than that reported to date for carbon black (typically smaller than 900 m²/g) or for carbon nanotubes (CNTs), from ≈100 to 1000 m²/g and is similar to activated carbon.

Structure

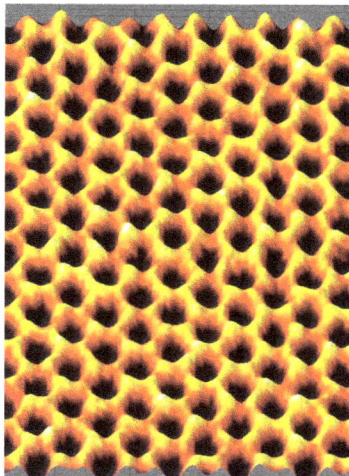

Scanning probe microscopy image of graphene.

Graphene is a crystalline allotrope of carbon with 2-dimensional properties. Its carbon atoms are densely packed in a regular atomic-scale chicken wire (hexagonal) pattern.

Each atom has four bonds, one σ bond with each of its three neighbors and one π-bond that is oriented out of plane. The atoms are about 1.42 Å apart.

Graphene's hexagonal lattice can be regarded as two interleaving triangular lattices. This perspective was successfully used to calculate the band structure for a single graphite layer using a tight-binding approximation.

Graphene's stability is due to its tightly packed carbon atoms and a sp² orbital hybridization – a combination of orbitals s, p_x and p_y that constitute the σ-bond. The final p_z electron makes up the π-bond. The π-bonds hybridize together to form the π-band and π*-bands. These bands are responsible for most of graphene's notable electronic properties, via the half-filled band that permits free-moving electrons. Recent quantitative estimates of aromatic stabilization and limiting size derived from the enthalpies of hydrogenation (ΔH_{hydro}) agree well with the literature reports.

Graphene sheets in solid form usually show evidence in diffraction for graphite's (002) layering. This is true of some single-walled nanostructures. However, unlayered graphene with only (hk0) rings has been found in the core of presolar graphite onions. TEM studies show faceting at defects in flat graphene sheets and suggest a role for two-dimensional crystallization from a melt.

Graphene can self-repair holes in its sheets, when exposed to molecules containing carbon, such as hydrocarbons. Bombarded with pure carbon atoms, the atoms perfectly align into hexagons, completely filling the holes.

The atomic structure of isolated, single-layer graphene was studied by transmission electron microscopy (TEM) on sheets of graphene suspended between bars of a metallic grid. Electron diffraction patterns showed the expected honeycomb lattice. Suspended graphene also showed "rippling" of the flat sheet, with amplitude of about one nanometer. These ripples may be intrinsic to the material as a result of the instability of two-dimensional crystals, or may originate from the ubiquitous dirt seen in all TEM images of graphene. Atomic resolution real-space images of isolated, single-layer graphene on SiO_2 substrates are available via scanning tunneling microscopy. Photoresist residue, which must be removed to obtain atomic-resolution images, may be the "adsorbates" observed in TEM images, and may explain the observed rippling. Rippling on SiO_2 is caused by conformation of graphene to the underlying SiO_2, and is not intrinsic.

Chemical

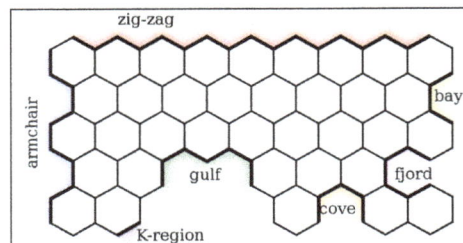

Names for graphene edge topologies.

Graphene is the only form of carbon (or solid material) in which every atom is available for chemical reaction from two sides (due to the 2D structure). Atoms at the edges of a graphene sheet have special chemical reactivity. Graphene has the highest ratio of edge atoms of any allotrope. Defects within a sheet increase its chemical reactivity. The onset temperature of reaction between the basal plane of single-layer graphene and oxygen gas is below 260 °C (530 K). Graphene burns at very low temperature (e.g., 350 °C (620 K)). Graphene is commonly modified with oxygen- and nitrogen-containing functional groups and analyzed by infrared spectroscopy and X-ray photoelectron spectroscopy. However, determination of structures of graphene with oxygen- and nitrogen- functional groups requires the structures to be well controlled.

In 2013, Stanford University physicists reported that single-layer graphene is a hundred times more chemically reactive than thicker sheets.

Electronic

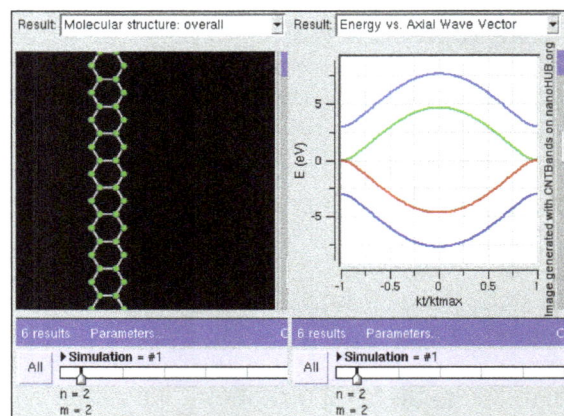

GNR band structure for zig-zag orientation. Tightbinding calculations show that zig-zag orientation is always metallic.

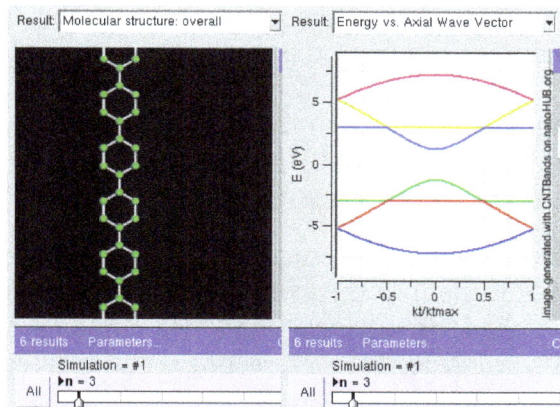

GNR band structure for armchair orientation. Tightbinding calculations show that armchair orientation can be semiconducting or metallic depending on width (chirality).

Graphene is a zero-gap semiconductor, because its conduction and valence bands meet at the Dirac points. The Dirac points are six locations in momentum space, on the edge of the Brillouin zone, divided into two non-equivalent sets of three points. The two sets are labeled K and K'. The sets give graphene a valley degeneracy of $gv = 2$. By contrast, for traditional semiconductors the primary point of interest is generally Γ, where momentum is zero. Four electronic properties separate it from other condensed matter systems.

However, if the in-plane direction is no longer infinite, but confined, its electronic structure would change. They are referred to as graphene nanoribbons. If it is "zig-zag", the bandgap would still be zero. If it is "armchair", the bandgap would be non-zero.

Electronic Spectrum

Electrons propagating through graphene's honeycomb lattice effectively lose their mass, producing quasi-particles that are described by a 2D analogue of the Dirac equation rather than the Schrödinger equation for spin-$\frac{1}{2}$ particles.

Dispersion Relation

When the atoms are placed onto the graphene hexagonal lattice, the overlap between the $p_z(\pi)$ orbitals and the s or the p_x and p_y orbitals is zero by symmetry. The p_z electrons forming the π bands in graphene can therefore be treated independently. Within this π-band approximation, using a conventional tight-binding model, the dispersion relation (restricted to first-nearest-neighbor interactions only) that produces energy of the electrons with wave vector k is:

$$E = \pm \sqrt{\gamma_0^2 \left(1 + 4\cos^2 \frac{k_y a}{2} + 4\cos \frac{k_y a}{2} \cdot \cos \frac{k_x \sqrt{3} a}{2} \right)}$$

with the nearest-neighbor (π orbitals) hopping energy $\gamma_0 \approx 2.8$ eV and the lattice constant $a \approx 2.46$ Å. The conduction and valence bands, respectively, correspond to the different signs. With one p_z electron per atom in this model the valence band is fully occupied, while the conduction band is vacant. The two bands touch at the zone corners (the K point in the Brillouin zone), where there is a zero density of states but no band gap. The graphene sheet thus displays a semimetallic (or zero-gap semiconductor) character, although the same cannot be said of a graphene sheet rolled into a carbon nanotube, due to its curvature. Two of the six Dirac points are independent, while the rest are equivalent by symmetry. In the vicinity of the K-points the energy depends *linearly* on the wave vector, similar to a relativistic particle. Since an elementary cell of the lattice has a basis of two atoms, the wave function has an effective 2-spinor structure.

As a consequence, at low energies, even neglecting the true spin, the electrons can be described by an equation that is formally equivalent to the massless Dirac equation. Hence, the electrons and holes are called Dirac fermions. This pseudo-relativistic description is restricted to the chiral limit, i.e., to vanishing rest mass M_0, which leads to interesting additional features:

$$v_F \, \vec{\sigma} \cdot \nabla \, \psi(\mathrm{r}) = E \psi(\mathrm{r}).$$

Here $v_F \sim 10^6$ m/s (.003 c) is the Fermi velocity in graphene, which replaces the velocity of light in the Dirac theory; $\vec{\sigma}$ is the vector of the Pauli matrices, $\psi(\mathrm{r})$ is the two-component wave function of the electrons, and E is their energy.

The equation describing the electrons' linear dispersion relation is:

$$E = \hbar v_F \sqrt{k_x^2 + k_y^2}$$

where the wavevector k is measured from the Dirac points (the zero of energy is chosen here to coincide with the Dirac points). The equation uses a pseudospin matrix formula that describes two sublattices of the honeycomb lattice.

Single-atom Wave Propagation

Electron waves in graphene propagate within a single-atom layer, making them sensitive to the proximity of other materials such as high-κ dielectrics, superconductors and ferromagnetics.

Electron Transport

Graphene displays remarkable electron mobility at room temperature, with reported values in excess of 15000 cm²·V⁻¹·s⁻¹. Hole and electron mobilities were expected to be nearly identical. The mobility is nearly independent of temperature between 10 K and 100 K, which implies that the dominant scattering mechanism is defect scattering. Scattering by graphene's acoustic phonons intrinsically limits room temperature mobility to 200000 cm²·V⁻¹·s⁻¹ at a carrier density of 10¹² cm⁻², 10×10⁶ times greater than copper.

The corresponding resistivity of graphene sheets would be 10⁻⁶ Ω·cm. This is less than the resistivity of silver, the lowest otherwise known at room temperature. However, on SiO₂ substrates, scattering of electrons by optical phonons of the substrate is a larger effect than scattering by graphene's own phonons. This limits mobility to 40000 cm²·V⁻¹·s⁻¹.

Charge transport has major concerns due to adsorption of contaminants such as water and oxygen molecules. This leads to non-repetitive and large hysteresis I-V characteristics. Researchers must carry out electrical measurements in vacuum. The protection of graphene surface by a coating with materials such as SiN, PMMA, h-BN, etc., have been discussed by researchers. In January 2015, the first stable graphene device operation in air over several weeks was reported, for graphene whose surface was protected by aluminum oxide. In 2015 lithium-coated graphene exhibited superconductivity, a first for graphene.

Electrical resistance in 40-nanometer-wide nanoribbons of epitaxial graphene changes in discrete steps. The ribbons' conductance exceeds predictions by a factor of 10. The ribbons can act more like optical waveguides or quantum dots, allowing electrons to flow smoothly along the ribbon edges. In copper, resistance increases in proportion to length as electrons encounter impurities.

Transport is dominated by two modes. One is ballistic and temperature independent, while the other is thermally activated. Ballistic electrons resemble those in cylindrical carbon nanotubes. At room temperature, resistance increases abruptly at a particular length—the ballistic mode at 16 micrometres and the other at 160 nanometres (1% of the former length).

Graphene electrons can cover micrometer distances without scattering, even at room temperature.

Despite zero carrier density near the Dirac points, graphene exhibits a minimum conductivity on the order of $4e^2 / (\pi h)$. The origin of this minimum conductivity is still unclear. However, rippling of the graphene sheet or ionized impurities in the SiO_2 substrate may lead to local puddles of carriers that allow conduction. Several theories suggest that the minimum conductivity should be $4e^2 / h$; however, most measurements are of order $4e^2 / h$ or greater and depend on impurity concentration.

Near zero carrier density graphene exhibits positive photoconductivity and negative photoconductivity at high carrier density. This is governed by the interplay between photoinduced changes of both the Drude weight and the carrier scattering rate.

Graphene doped with various gaseous species (both acceptors and donors) can be returned to an undoped state by gentle heating in vacuum. Even for dopant concentrations in excess of 10^{12} cm^{-2} carrier mobility exhibits no observable change. Graphene doped with potassium in ultra-high vacuum at low temperature can reduce mobility 20-fold. The mobility reduction is reversible on heating the graphene to remove the potassium.

Due to graphene's two dimensions, charge fractionalization (where the apparent charge of individual pseudoparticles in low-dimensional systems is less than a single quantum) is thought to occur. It may therefore be a suitable material for constructing quantum computers using anyonic circuits.

Anomalous Quantum Hall effect

The quantum Hall effect is a quantum mechanical version of the Hall effect, which is the production of transverse (perpendicular to the main current) conductivity in the presence of a magnetic field. The quantization of the Hall effect σ_{xy} at integer multiples (the "Landau level") of the basic quantity e^2 / h (where e is the elementary electric charge and h is Planck's constant) It can usually be observed only in very clean silicon or gallium arsenide solids at temperatures around 3 K and very high magnetic fields.

Graphene shows the quantum Hall effect with respect to conductivity quantization: the effect is *anomalous* in that the sequence of steps is shifted by 1/2 with respect to the standard sequence and with an additional factor of 4. Graphene's Hall conductivity is $\sigma_{xy} = \pm 4 \left(N + 1/2 \right) e^2 / h$, where N is the Landau level and the double valley and double spin degeneracies give the factor of 4. These anomalies are present at room temperature, i.e. at roughly 20 °C (293 K).

This behavior is a direct result of graphene's massless Dirac electrons. In a magnetic field, their spectrum has a Landau level with energy precisely at the Dirac point. This level is a consequence of the Atiyah–Singer index theorem and is half-filled in neutral graphene, leading to the "+1/2" in the Hall conductivity. Bilayer graphene also shows the quantum Hall effect, but with only one of the two anomalies (i.e. $\sigma_{xy} = \pm 4 \cdot N \cdot e^2 / h$). In the second anomaly, the first plateau at $N=0$ is absent, indicating that bilayer graphene stays metallic at the neutrality point.

Unlike normal metals, graphene's longitudinal resistance shows maxima rather than minima for integral values of the Landau filling factor in measurements of the Shubnikov–de Haas oscillations, whereby the term *integral* quantum Hall effect. These oscillations show a phase shift of π, known as Berry's phase. Berry's phase arises due to the zero effective carrier mass near the Dirac points. The temperature dependence of the oscillations reveals that the carriers have a non-zero cyclotron mass, despite their zero effective mass.

Graphene samples prepared on nickel films, and on both the silicon face and carbon face of silicon carbide, show the anomalous effect directly in electrical measurements. Graphitic layers on the carbon face of silicon carbide show a clear Dirac spectrum in angle-resolved photoemission experiments, and the effect is observed in cyclotron resonance and tunneling experiments.

Strong Magnetic Fields

In magnetic fields above 10 tesla or so additional plateaus of the Hall conductivity at $\sigma_{xy} = v e^2 / h$ with v = 0, ±1, ±4 are observed. A plateau at v = 3 and the fractional quantum Hall effect at $v = \frac{1}{3}$ were also reported.

These observations with v = 0, ±1, ±3, ±4 indicate that the four-fold degeneracy (two valley and two spin degrees of freedom) of the Landau energy levels is partially or completely lifted.

Casimir Effect

The Casimir effect is an interaction between disjoint neutral bodies provoked by the fluctuations of the electrodynamical vacuum. Mathematically it can be explained by considering the normal modes of electromagnetic fields, which explicitly depend on the boundary (or matching) conditions on the interacting bodies' surfaces. Since graphene/electromagnetic field interaction is strong for a one-atom-thick material, the Casimir effect is of growing interest.

Van der Waals Force

The Van der Waals force (or dispersion force) is also unusual, obeying an inverse cubic, asymptotic power law in contrast to the usual inverse quartic.

'Massive' Electrons

Graphene's unit cell has two identical carbon atoms and two zero-energy states: one in which the electron resides on atom A, the other in which the electron resides on atom B. However, if the two atoms in the unit cell are not identical, the situation changes. Hunt et al. show that placing hexagonal boron nitride (h-BN) in contact with graphene can alter the potential felt at atom A versus atom B enough that the electrons develop a mass and accompanying band gap of about 30 meV [0.03 Electron Volt(eV)].

The mass can be positive or negative. An arrangement that slightly raises the energy of an electron on atom A relative to atom B gives it a positive mass, while an arrangement that raises the energy of atom B produces a negative electron mass. The two versions behave alike and are indistinguishable via optical spectroscopy. An electron traveling from a positive-mass region to a negative-mass region must cross an intermediate region where its mass once again becomes zero. This region is gapless and therefore metallic. Metallic modes bounding semiconducting regions of opposite-sign mass is a hallmark of a topological phase and display much the same physics as topological insulators.

If the mass in graphene can be controlled, electrons can be confined to massless regions by surrounding them with massive regions, allowing the patterning of quantum dots, wires, and other mesoscopic structures. It also produces one-dimensional conductors along the boundary. These wires would be protected against backscattering and could carry currents without dissipation.

Optical

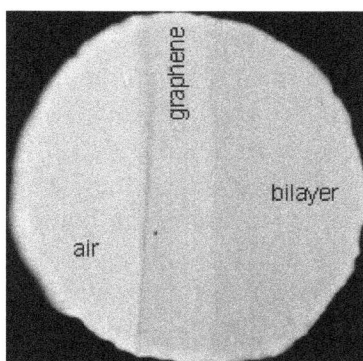

Photograph of graphene in transmitted light. This one-atom-thick crystal can be seen with the naked eye because it absorbs approximately 2.6% of green light, and 2.3% of red light.

Graphene's unique optical properties produce an unexpectedly high opacity for an atomic monolayer in vacuum, absorbing $\pi\alpha \approx 2.3\%$ of red light, where α is the fine-structure constant. This is a consequence of the "unusual low-energy electronic structure of monolayer graphene that features electron and hole conical bands meeting each other at the Dirac point [which] is qualitatively different from more common quadratic massive bands". Based on the Slonczewski–Weiss–McClure (SWMcC) band model of graphite, the interatomic distance, hopping value and frequency cancel when optical conductance is calculated using Fresnel equations in the thin-film limit.

Although confirmed experimentally, the measurement is not precise enough to improve on other techniques for determining the fine-structure constant.

Multi-Parametric Surface Plasmon Resonance was used to characterize both thickness and refractive index of chemical-vapor-deposition (CVD)-grown graphene films. The measured refractive index and extinction coefficient values at 670 nm wavelength are 3.135 and 0.897, respectively. The thickness was determined as 3.7Å from a 0.5mm area, which agrees with 3.35Å reported for layer-to-layer carbon atom distance of graphite crystals. The method can be further used also for real-time label-free interactions of graphene with organic and inorganic substances. Furthermore, the existence of unidirectional surface plasmons in the nonreciprocal graphene-based gyrotropic interfaces has been demonstrated theoretically. By efficiently controlling the chemical potential of graphene, the unidirectional working frequency can be continuously tunable from THz to near-infrared and even visible. Particularly, the unidirectional frequency bandwidth can be 1– 2 orders of magnitude larger than that in metal under the same magnetic field, which arises from the superiority of extremely small effective electron mass in graphene.

Graphene's band gap can be tuned from 0 to 0.25 eV (about 5 micrometre wavelength) by applying voltage to a dual-gate bilayer graphene field-effect transistor (FET) at room temperature. The optical response of graphene nanoribbons is tunable into the terahertz regime by an applied magnetic field. Graphene/graphene oxide systems exhibit electrochromic behavior, allowing tuning of both linear and ultrafast optical properties.

A graphene-based Bragg grating (one-dimensional photonic crystal) has been fabricated and demonstrated its capability for excitation of surface electromagnetic waves in the periodic structure by using 633 nm He–Ne laser as the light source.

Saturable Absorption

Such unique absorption could become saturated when the input optical intensity is above a threshold value. This nonlinear optical behavior is termed saturable absorption and the threshold value is called the saturation fluence. Graphene can be saturated readily under strong excitation over the visible to near-infrared region, due to the universal optical absorption and zero band gap. This has relevance for the mode locking of fiber lasers, where fullband mode locking has been achieved by graphene-based saturable absorber. Due to this special property, graphene has wide application in ultrafast photonics. Moreover, the optical response of graphene/graphene oxide layers can be tuned electrically. Saturable absorption in graphene could occur at the Microwave and Terahertz band, owing to its wideband optical absorption property. The microwave saturable absorption in graphene demonstrates the possibility of graphene microwave and terahertz photonics devices, such as a microwave saturable absorber, modulator, polarizer, microwave signal processing and broad-band wireless access networks.

Nonlinear Kerr Effect

Under more intensive laser illumination, graphene could also possess a nonlinear phase shift due to the optical nonlinear Kerr effect. Based on a typical open and close aperture z-scan measurement, graphene possesses a giant nonlinear Kerr coefficient of 10^{-7} cm^2·W^{-1}, almost nine orders of magnitude larger than that of bulk dielectrics. This suggests that graphene may be a powerful nonlinear Kerr medium, with the possibility of observing a variety of nonlinear effects, the most important of which is the soliton.

Excitonic

First-principle calculations with quasiparticle corrections and many-body effects are performed to study the electronic and optical properties of graphene-based materials. The approach is described as three stages. With GW calculation, the properties of graphene-based materials are accurately investigated, including bulk graphene, nanoribbons, edge and surface functionalized armchair oribbons, hydrogen saturated armchair ribbons, Josephson effect in graphene SNS junctions with single localized defect and armchair ribbon scaling properties.

Stability

Ab initio calculations show that a graphene sheet is thermodynamically unstable if its size is less than about 20 nm ("graphene is the least stable structure until about 6000 atoms") and becomes the most stable fullerene (as within graphite) only for molecules larger than 24,000 atoms.

Thermal Conductivity

Thermal transport in graphene is an active area of research, which has attracted attention because of the potential for thermal management applications. Early measurements of the thermal conductivity of suspended graphene reported an exceptionally large thermal conductivity of approximately 5300 $W \cdot m^{-1} \cdot K^{-1}$, compared with the thermal conductivity of pyrolytic graphite of approximately 2000 $W \cdot m^{-1} \cdot K^{-1}$ at room temperature. However, later studies have questioned whether this ultrahigh value had been overestimated, and have instead measured a wide range of thermal conductivities between 1500 – 2500 $W \cdot m^{-1} \cdot K^{-1}$ for suspended single layer graphene. The large range in the reported thermal conductivity can be caused by large measurement uncertainties as well as variations in the graphene quality and processing conditions. In addition, it is known that when single-layer graphene is supported on an amorphous material, the thermal conductivity is reduced to about 500 – 600 $W \cdot m^{-1} \cdot K^{-1}$ at room temperature as a result of scattering of graphene lattice waves by the substrate, and can be even lower for few layer graphene encased in amorphous oxide. Likewise, polymeric residue can contribute to a similar decrease in the thermal conductivity of suspended graphene to approximately 500 – 600 $W \cdot m^{-1} \cdot K^{-1}$ for bilayer graphene.

It has been suggested that the isotopic composition, the ratio of ^{12}C to ^{13}C, has a significant impact on the thermal conductivity. For example, isotopically pure ^{12}C graphene has higher thermal conductivity than either a 50:50 isotope ratio or the naturally occurring 99:1 ratio. It can be shown by using the Wiedemann–Franz law, that the thermal conduction is phonon-dominated. However, for a gated graphene strip, an applied gate bias causing a Fermi energy shift much larger than $k_B T$ can cause the electronic contribution to increase and dominate over the phonon contribution at low temperatures. The ballistic thermal conductance of graphene is isotropic.

Potential for this high conductivity can be seen by considering graphite, a 3D version of graphene that has basal plane thermal conductivity of over a 1000 $W \cdot m^{-1} \cdot K^{-1}$ (comparable to diamond). In graphite, the c-axis (out of plane) thermal conductivity is over a factor of ~100 smaller due to the weak binding forces between basal planes as well as the larger lattice spacing. In addition, the ballistic thermal conductance of graphene is shown to give the lower limit of the ballistic thermal conductances, per unit circumference, length of carbon nanotubes.

Despite its 2-D nature, graphene has 3 acoustic phonon modes. The two in-plane modes (LA, TA) have a linear dispersion relation, whereas the out of plane mode (ZA) has a quadratic dispersion relation. Due to this, the T^2 dependent thermal conductivity contribution of the linear modes is dominated at low temperatures by the $T^{1.5}$ contribution of the out of plane mode. Some graphene phonon bands display negative Grüneisen parameters. At low temperatures (where most optical modes with positive Grüneisen parameters are still not excited) the contribution from the negative Grüneisen parameters will be dominant and thermal expansion coefficient (which is directly proportional to Grüneisen parameters) negative. The lowest negative Grüneisen parameters correspond to the lowest transverse acoustic ZA modes. Phonon frequencies for such modes increase with the in-plane lattice parameter since atoms in the layer upon stretching will be less free to move in the z direction. This is similar to the behavior of a string, which, when it is stretched, will have vibrations of smaller amplitude and higher frequency. This phenomenon, named "membrane effect," was predicted by Lifshitz in 1952.

Mechanical

The carbon–carbon bond length in graphene is about 0.142 nanometers. Graphene sheets stack to form graphite with an interplanar spacing of 0.335 nm.

Graphene is the strongest material ever tested, with an intrinsic tensile strength of 130 GPa and a Young's modulus (stiffness) of 1 TPa (150000000 psi). The Nobel announcement illustrated this by saying that a 1 square meter graphene hammock would support a 4 kg cat but would weigh only as much as one of the cat's whiskers, at 0.77 mg (about 0.001% of the weight of 1 m² of paper).

Large-angle-bent graphene monolayer has been achieved with negligible strain, showing mechanical robustness of the two-dimensional carbon nanostructure. Even with extreme deformation, excellent carrier mobility in monolayer graphene can be preserved.

The spring constant of suspended graphene sheets has been measured using an atomic force microscope (AFM). Graphene sheets were suspended over SiO_2 cavities where an AFM tip was used to apply a stress to the sheet to test its mechanical properties. Its spring constant was in the range 1–5 N/m and the stiffness was 0.5 TPa, which differs from that of bulk graphite. These intrinsic properties could lead to applications such as NEMS as pressure sensors and resonators. Due to its large surface energy and out of plane ductility, flat graphene sheets are unstable with respect to scrolling, i.e. bending into a cylindrical shape, which is its lower-energy state.

As is true of all materials, regions of graphene are subject to thermal and quantum fluctuations in relative displacement. Although the amplitude of these fluctuations is bounded in 3D structures (even in the limit of infinite size), the Mermin–Wagner theorem shows that the amplitude of long-wavelength fluctuations grows logarithmically with the scale of a 2D structure, and would therefore be unbounded in structures of infinite size. Local deformation and elastic strain are negligibly affected by this long-range divergence in relative displacement. It is believed that a sufficiently large 2D structure, in the absence of applied lateral tension, will bend and crumple to form a fluctuating 3D structure. Researchers have observed ripples in suspended layers of graphene, and it has been proposed that the ripples are caused by thermal fluctuations in the material. As a consequence of these dynamical deformations, it is debatable whether graphene is truly a 2D structure. It has recently been shown that these ripples, if amplified through the introduction of vacancy defects, can impart a negative Poisson's ratio into graphene, resulting in the thinnest auxetic material known so far.

Graphene nanosheets have been incorporated into a Ni matrix through a plating process to form Ni-graphene composites on a target substrate. The enhancement in mechanical properties of the composites is attributed to the high interaction between Ni and graphene and the prevention of the dislocation sliding in the Ni matrix by the graphene.

Fracture Toughness

In 2014, researchers from Rice University and the Georgia Institute of Technology have indicated that despite its strength, graphene is also relatively brittle, with a fracture toughness of about 4 MPa√m. This indicates that imperfect graphene is likely to crack in a brittle manner like ceramic materials, as opposed to many metallic materials which tend to have fracture toughnesses in the range of 15–50 MPa√m. Later in 2014, the Rice team announced that graphene showed a greater ability to distribute force from an impact than any known material, ten times that of steel per unit weight. The force was transmitted at 22.2 kilometres per second (13.8 mi/s).

Mechanical Properties of Polycrystalline Graphene

Various methods – most notably, chemical vapor deposition (CVD), have been developed to produce large-scale graphene needed for device applications. Such methods often synthesize polycrystalline graphene. The mechanical properties of polycrystalline graphene is affected by the nature of the defects, such as grain-boundaries (GB) and vacancies, present in the system and the average grain-size. How the mechanical properties change with such defects have been investigated by researchers, theoretically and experimentally.

Graphene grain boundaries typically contain heptagon-pentagon pairs. The arrangement of such defects depends on whether the GB is in zig-zag or armchair direction. It further depends on the tilt-angle of the GB. In 2010, researchers from Brown University computationally predicted that as the tilt-angle increases, the grain boundary strength also increases.They showed that the weakest link in the grain boundary is at the critical bonds of the heptagon rings. As the grain boundary angle increases, the strain in these heptagon rings decreases, causing the grain-boundary to be stronger than lower-angle GBs. They proposed that, in fact, for sufficiently large angle GB, the strength of the GB is similar to pristine graphene. In 2012, it was further shown that the strength can increase or decrease, depending on the detailed arrangements of the defects. These predictions have since been supported by experimental evidences. In a 2013 study led by James Hone's group, researchers probed the elastic stiffness and strength of CVD-grown graphene by combining nano-indentation and high-resolution TEM. They found that the elastic stiffness is identical and strength is only slightly lower than those in pristine graphene. In the same year, researchers from UC Berkeley and UCLA probed bi-crystalline graphene with TEM and AFM. They found that the strength of grain-boundaries indeed tend to increase with the tilt angle.

While the presence of vacancies is not only prevalent in polycrystalline graphene, vacancies can have significant effects on the strength of graphene. The general consensus is that the strength decreases along with increasing densities of vacancies. In fact, various studies have shown that for graphene with for sufficiently low density of vacancies, the strength does not vary significantly from that of pristine graphene. On the other hand, high density of vacancies can severely reduce the strength of graphene.

Compared to the fairly well-understood nature of the effect that grain boundary and vacancies have on the mechanical properties of graphene, there is no clear consensus on the general effect that the average grain size has on the strength of polycrystalline graphene. In fact, three notable theoretical/computational studies on this topic have led to three different conclusions. First, in 2012, Kotakoski and Myer studied the mechanical properties of polycrystalline graphene with "realistic atomistic model", using molecular-dynamics (MD) simulation. To emulate the growth mechanism of CVD, they first randomly selected nucleation sites that are at least 5A (arbitrarily chosen) apart from other sites. Polycrystalline graphene was generated from these nucleation sites and was subsequently annealed at 3000K, then quenched. Based on this model, they found that cracks are initiated at grain-boundary junctions, but the grain size does not significantly affect the strength. Second, in 2013, Z. Song et al. used MD simulations to study the mechanical properties of polycrystalline graphene with uniform-sized hexagon-shaped grains. The hexagon grains were oriented in various lattice directions and the GBs consisted of only heptagon, pentagon, and hexagonal carbon rings. The motivation behind such model was that similar systems had been experimentally observed in graphene flakes grown on the surface of liquid copper. While they also noted that crack is typically initiated at the triple junctions, they found that as the grain size decreases, the yield strength of graphene increases. Based on this finding, they proposed that polycrystalline follows pseudo Hall-Petch relationship. Third, in 2013, Z. D. Sha et al. studied the effect of grain size on the properties of polycrystalline graphene, by modelling the grain patches using Voronoi construction. The GBs in this model consisted of heptagon, pentagon, and hexagon, as well as squares, octagons, and vacancies. Through MD simulation, contrary to the fore-mentioned study, they found inverse Hall-Petch relationship, where the strength of graphene increases as the grain size increases. Experimental observations and other theoretical predictions also gave differing conclusions, similar to the three given above. Such discrepencies show the complexity of the effects that grain size, arrangements of defects, and the nature of defects have on the mechanical properties of polycrystalline graphene.

Spin Transport

Graphene is claimed to be an ideal material for spintronics due to its small spin-orbit interaction and the near absence of nuclear magnetic moments in carbon (as well as a weak hyperfine interaction). Electrical spin current injection and detection has been demonstrated up to room temperature. Spin coherence length above 1 micrometre at room temperature was observed, and control of the spin current polarity with an electrical gate was observed at low temperature.

Strong Magnetic Fields

Graphene's quantum Hall effect in magnetic fields above 10 Teslas or so reveals additional interesting features. Additional plateaus of the Hall conductivity at $\sigma_{xy} = \nu e^2 / h$ with $\nu = 0, \pm 1, \pm 4$ are observed. Also, the observation of a plateau at $\nu = 3$ and the fractional quantum Hall effect at $\nu = 1/3$ were reported.

These observations with $\nu = 0, \pm 1, \pm 3, \pm 4$ indicate that the four-fold degeneracy (two valley and two spin degrees of freedom) of the Landau energy levels is partially or completely lifted. One hypothesis is that the magnetic catalysis of symmetry breaking is responsible for lifting the degeneracy.

Spintronic and magnetic properties can be present in graphene simultaneously. Low-defect graphene nanomeshes manufactured by using a non-lithographic method exhibit large-amplitude ferromagnetism even at room temperature. Additionally a spin pumping effect is found for fields applied in parallel with the planes of few-layer ferromagnetic nanomeshes, while a magnetoresistance hysteresis loop is observed under perpendicular fields.

Magnetic

In 2014 researchers magnetized graphene by placing it on an atomically smooth layer of magnetic yttrium iron garnet. The graphene's electronic properties were unaffected. Prior approaches involved doping graphene with other substances. The dopant's presence negatively affected its electronic properties.

Biological

Researchers at the Graphene Research Centre at the National University of Singapore (NUS) discovered in 2011 the amazing ability of graphene to accelerate the osteogenic differentiation of human Mesenchymal Stem Cells without the use of biochemical inducers.

In 2015 researchers used graphene to create sensitive biosensors by using epitaxial graphene on silicon carbide. The sensors bind to the 8-hydroxydeoxyguanosine (8-OHdG) and is capable of selective binding with antibodies. The presence of 8-OHdG in blood, urine and saliva is commonly associated with DNA damage. Elevated levels of 8-OHdG have been linked to increased risk of developing several cancers.

The Cambridge Graphene Centre and the University of Trieste in Italy conducted a collaborative research on use of Graphene as electrodes to interact with brain neurons. The research was recently published in the journal of ACS Nano.

The research revealed that uncoated Graphene can be used as neuro-interface electrode without altering or damaging the neural functions such as signal loss or formation of scar tissue. Graphene electrodes in body stay significantly more stable than modern day electrodes (of tungsten or silicon) because of its unique properties such as flexibility, bio-compatibility, and conductivity. It could possibly help in restoring sensory function or motor disorders in paralysis or Parkinson patients.

Support Substrate

The electronics property of graphene can be significantly influenced by the supporting substrate. Studies of graphene monolayers on clean and hydrogen(H)-passivated silicon (100) (Si(100)/H) surfaces have been performed. The Si(100)/H surface does not perturb the electronic properties of graphene, whereas the interaction between the clean Si(100) surface and graphene changes the electronic states of graphene significantly. This effect results from the covalent bonding between C and surface Si atoms, modifying the π-orbital network of the graphene layer. The local density of states shows that the bonded C and Si surface states are highly disturbed near the Fermi energy.

Forms

Monolayer Sheets

In 2013 a group of Polish scientists presented a production unit that allows the manufacture of continuous monolayer sheets. The process is based on graphene growth on a liquid metal matrix. The product of this process was called HSMG.

Bilayer Graphene

Bilayer graphene displays the anomalous quantum Hall effect, a tunable band gap and potential for excitonic condensation –making it a promising candidate for optoelectronic and nanoelectronic applications. Bilayer graphene typically can be found either in twisted configurations where the two layers are rotated relative to each other or graphitic Bernal stacked configurations where half the atoms in one layer lie atop half the atoms in the other. Stacking order and orientation govern the optical and electronic properties of bilayer graphene.

One way to synthesize bilayer graphene is via chemical vapor deposition, which can produce large bilayer regions that almost exclusively conform to a Bernal stack geometry.

Graphene Superlattices

Periodically stacked graphene and its insulating isomorph provide a fascinating structural element in implementing highly functional superlattices at the atomic scale, which offers possibilities in designing nanoelectronic and photonic devices. Various types of superlattices can be obtained by stacking graphene and its related forms. The energy band in layer-stacked superlattices is found to be more sensitive to the barrier width than that in conventional III–V semiconductor superlattices. When adding more than one atomic layer to the barrier in each period, the coupling of electronic wavefunctions in neighboring potential wells can be significantly reduced, which leads to the degeneration of continuous subbands into quantized energy levels. When varying the well width, the energy levels in the potential wells along the L-M direction behave distinctly from those along the K-H direction.

Graphene Nanoribbons

Graphene nanoribbons ("nanostripes" in the "zig-zag" orientation), at low temperatures, show spin-polarized metallic edge currents, which also suggests applications in the new field of spintronics. In the "armchair" orientation, the edges behave like semiconductors.

Graphene Quantum Dots

Graphene quantum dots (GQDs) have mainly been fabricated by the microwave assisted hydrothermal method (MAH), the Soft-Template method, the hydrothermal method, the ultrasonic exfoliation method, the electron beam lithography method, the chemical synthesis method, the electrochemical preparation method, the graphene oxide (GO) reduction method, and the C60 catalytic transformation method, etc.

Graphene Oxide

Using paper-making techniques on dispersed, oxidized and chemically processed graphite in water, the monolayer flakes form a single sheet and create strong bonds. These sheets, called graphene oxide paper, have a measured tensile modulus of 32 GPa. The chemical property of graphite oxide is related to the functional groups attached to graphene sheets. These can change the polymerization pathway and similar chemical processes. Graphene oxide flakes in polymers display enhanced photo-conducting properties. Graphene is normally hydrophobic and impermeable to all gases and liquids (vacuum-tight). However, when formed into graphene oxide-based capillary membrane, both liquid water and water vapor flow through as quickly as if the membrane was not present.

Chemical Modification

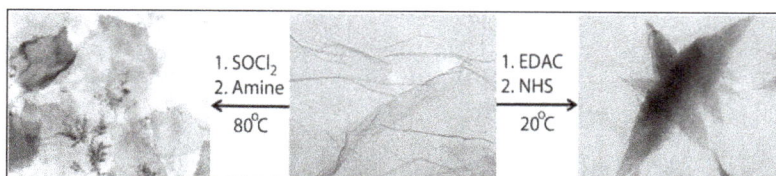

Photograph of single-layer graphene oxide undergoing high temperature chemical treatment, resulting in sheet folding and loss of carboxylic functionality, or through room temperature carbodiimide treatment, collapsing into star-like clusters.

Soluble fragments of graphene can be prepared in the laboratory through chemical modification of graphite. First, microcrystalline graphite is treated with an acidic mixture of sulfuric acid and nitric acid. A series of oxidation and exfoliation steps produce small graphene plates with carboxyl groups at their edges. These are converted to acid chloride groups by treatment with thionyl chloride; next, they are converted to the corresponding graphene amide via treatment with octadecylamine. The resulting material (circular graphene layers of 5.3 angstrom thickness) is soluble in tetrahydrofuran, tetrachloromethane and dichloroethane.

Refluxing single-layer graphene oxide (SLGO) in solvents leads to size reduction and folding of individual sheets as well as loss of carboxylic group functionality, by up to 20%, indicating thermal instabilities of SLGO sheets dependent on their preparation methodology. When using thionyl chloride, acyl chloride groups result, which can then form aliphatic and aromatic amides with a reactivity conversion of around 70–80%.

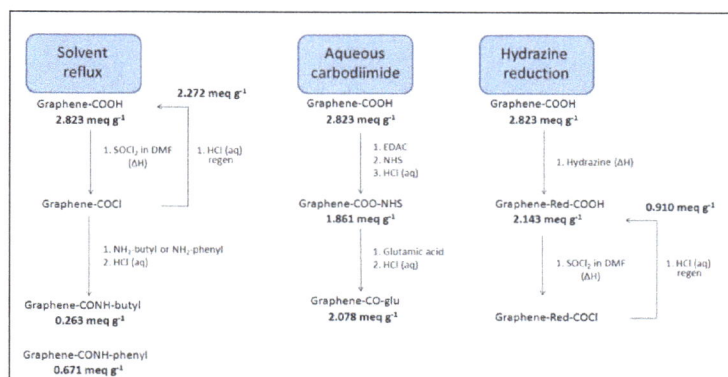

Boehm titration results for various chemical reactions of single-layer graphene oxide, which reveal reactivity of the carboxylic groups and the resultant stability of the SLGO sheets after treatment.

Hydrazine reflux is commonly used for reducing SLGO to SLG(R), but titrations show that only around 20–30% of the carboxylic groups are lost, leaving a significant number available for chemical attachment. Analysis of SLG(R) generated by this route reveals that the system is unstable and using a room temperature stirring with HCl (< 1.0 M) leads to around 60% loss of COOH functionality. Room temperature treatment of SLGO with carbodiimides leads to the collapse of the individual sheets into star-like clusters that exhibited poor subsequent reactivity with amines (c. 3–5% conversion of the intermediate to the final amide). It is apparent that conventional chemical treatment of carboxylic groups on SLGO generates morphological changes of individual sheets that leads to a reduction in chemical reactivity, which may potentially limit their use in composite synthesis. Therefore, chemical reactions types have been explored. SLGO has also been grafted with polyallylamine, cross-linked through epoxy groups. When filtered into graphene oxide paper, these composites exhibit increased stiffness and strength relative to unmodified graphene oxide paper.

Full hydrogenation from both sides of graphene sheet results in graphane, but partial hydrogenation leads to hydrogenated graphene. Similarly, both-side fluorination of graphene (or chemical and mechanical exfoliation of graphite fluoride) leads to fluorographene (graphene fluoride), while partial fluorination (generally halogenation) provides fluorinated (halogenated) graphene.

Graphene Ligand/Complex

Graphene can be a ligand to coordinate metals and metal ions by introducing functional groups. Structures of graphene ligands are similar to e.g. metal-porphyrin complex, metal-phthalocyanine complex, and metal-phenanthroline complex. Copper and nickel ions can be coordinated with graphene ligands.

Graphene Fiber

In 2011, researchers reported a novel yet simple approach to fabricate graphene fibers from chemical vapor deposition grown graphene films. The method was scalable and controllable, delivering tunable morphology and pore structure by controlling the evaporation of solvents with suitable surface tension. Flexible all-solid-state supercapacitors based on this graphene fibers were demonstrated in 2013.

In 2015 intercalating small graphene fragments into the gaps formed by larger, coiled graphene sheets, after annealing provided pathways for conduction, while the fragments helped reinforce the fibers. The resulting fibers offered better thermal and electrical conductivity and mechanical strength. Thermal conductivity reached 1290 watts per meter per kelvin, while tensile strength reached 1080 megapascals.

In 2016, Kilometer-scale continuous graphene fibers with outstanding mechanical properties and excellent electrical conductivity are produced by high-throughput wet-spinning of graphene oxide liquid crystals followed by graphitization through a full-scale synergetic defect-engineering strategy. The graphene fibers with superior performances promise wide applications in functional textiles, lightweight motors, microelectronic devices, etc.

Tsinghua University in Beijing, led by Wei Fei of the Department of Chemical Engineering, claims to be able to create a carbon nanotube fibre which has a tensile strength of 80 gigapascals.

3D Graphene

In 2013, a three-dimensional honeycomb of hexagonally arranged carbon was termed 3D graphene. Recently, self-supporting 3D graphene has also been produced. 3D structures of graphene can be fabricated by using either CVD or solution based methods. A recent review by Khurram and Xu et al., have provided the summary of the state-of-the-art techniques for fabrication of the 3D structure of graphene and other related two-dimensional materials. Recently, researchers at Stony Brook University have reported a novel radical-initiated crosslinking method to fabricate porous 3D free-standing architectures of graphene and carbon nanotubes using nanomaterials as building blocks without any polymer matrix as support. These 3D graphene (all-carbon) scaffolds/foams have applications in several fields such as energy storage, filtration, thermal management and bio-medical devices and implants.

Box-shaped graphene (BSG) nanostructure appeared after mechanical cleavage of pyrolytic graphite has been reported recently. The discovered nanostructure is a multilayer system of parallel hollow nanochannels located along the surface and having quadrangular cross-section. The thickness of the channel walls is approximately equal to 1 nm. Potential fields of BSG application include: ultra-sensitive detectors, high-performance catalytic cells, nanochannels for DNA sequencing and manipulation, high-performance heat sinking surfaces, rechargeable batteries of enhanced performance, nanomechanical resonators, electron multiplication channels in emission nanoelectronic devices, high-capacity sorbents for safe hydrogen storage.

Three dimensional bilayer graphene has also been reported.

Pillared Graphene

Pillared graphene is a hybrid carbon, structure consisting of an oriented array of carbon nanotubes connected at each end to a sheet of graphene. It was first described theoretically by George Froudakis and colleagues of the University of Crete in Greece in 2008. Pillared graphene has not yet been synthesised in the laboratory, but it has been suggested that it may have useful electronic properties, or as a hydrogen storage material.

Reinforced Graphene

Graphene reinforced with embedded carbon nanotube reinforcing bars ("rebar") is easier to manipulate, while improving the electrical and mechanical qualities of both materials.

Functionalized single- or multiwalled carbon nanotubes are spin-coated on copper foils and then heated and cooled, using the nanotubes themselves as the carbon source. Under heating, the functional carbon groups decompose into graphene, while the nanotubes partially split and form in-plane covalent bonds with the graphene, adding strength. $\pi-\pi$ stacking domains add more strength. The nanotubes can overlap, making the material a better conductor than standard CVD-grown graphene. The nanotubes effectively bridge the grain boundaries found in conventional graphene. The technique eliminates the traces of substrate on which later-separated sheets were deposited using epitaxy.

Stacks of a few layers have been proposed as a cost-effective and physically flexible replacement for indium tin oxide (ITO) used in displays and photovoltaic cells.

Molded Graphene

A film of graphene that had been soaked in solvent to make it swell and become malleable was overlaid on an underlying substrate "former". The solvent evaporated over time, leaving behind a layer of graphene that had taken on the shape of the underlying structure. In this way the team was able to produce a range of relatively intricate micro-structured shapes. Features vary from 3.5 to 50 μm. Pure graphene and gold-decorated graphene were each successfully integrated with the substrate.

Graphene Aerogel

An aerogel made of graphene layers separated by carbon nanotubes was measured at 0.16 milligrams per cubic centimeter. A solution of graphene and carbon nanotubes in a mold is freeze dried to dehydrate the solution, leaving the aerogel. The material has superior elasticity and absorption. It can recover completely after more than 90% compression, and absorb up to 900 times its weight in oil, at a rate of 68.8 grams per second.

Graphene Nanocoil

In 2015 a coiled form of graphene was discovered in graphitic carbon (coal). The spiraling effect is produced by defects in the material's hexagonal grid that causes it to spiral along its edge, mimicking a Riemann surface, with the graphene surface approximately perpendicular to the axis. When voltage is applied to such a coil, current flows around the spiral, producing a magnetic field. The phenomenon applies to spirals with either zigzag or armchair patterns, although with different current distributions. Computer simulations indicated that a conventional spiral inductor of 205 microns in diameter could be matched by a nanocoil just 70 nanometers wide, with a field strength reaching as much as 1 tesla.

The nano-solenoids analyzed through computer models at Rice should be capable of producing powerful magnetic fields of about 1 tesla, about the same as the coils found in typical loudspeakers, according to Yakobson and his team – and about the same field strength as some MRI machines. They found the magnetic field would be strongest in the hollow, nanometer-wide cavity at the spiral's center.

A solenoid made with such a coil behaves as a quantum conductor whose current distribution between the core and exterior varies with applied voltage, resulting in nonlinear inductance.

Production

A rapidly increasing list of production techniques have been developed to enable graphene's use in commercial applications.

Isolated 2D crystals cannot be grown via chemical synthesis beyond small sizes even in principle, because the rapid growth of phonon density with increasing lateral size forces 2D crystallites to bend into the third dimension.

In all cases, graphene must bond to a substrate to retain its two-dimensional shape.

Exfoliation

As of 2014, exfoliation produced graphene with the lowest number of defects and highest electron mobility.

Geim and Novoselov initially used adhesive tape to pull graphene sheets away from graphite. Achieving single layers typically requires multiple exfoliation steps. After exfoliation the flakes are deposited on a silicon wafer. Crystallites larger than 1 mm and visible to the naked eye can be obtained.

Alternatively a sharp single-crystal diamond wedge penetrates onto the graphite source to cleave layers.

P. Boehm reported producing monolayer flakes of reduced graphene oxide in 1962. Rapid heating of graphite oxide and exfoliation yields highly dispersed carbon powder with a few percent of graphene flakes.

Another method is reduction of graphite oxide monolayer films, e.g. by hydrazine with annealing in argon/hydrogen with an almost intact carbon framework that allows efficient removal of functional groups. Measured charge carrier mobility exceeded 1,000 centimetres (393.70 in)/Vs.

In 2014 defect-free, unoxidized graphene-containing liquids were made from graphite using mixers that produce local shear rates greater than 10×10^4.

Burning a graphite oxide coated DVD produced a conductive graphene film (1738 siemens per meter) and specific surface area (1520 square meters per gram) that was highly resistant and malleable.

Dispersing graphite in a liquid medium can produce graphene by sonication followed by centrifugation, producing concentrations 2.1 mg/ml in N-methylpyrrolidone. Using a suitable ionic liquid as the dispersing liquid medium produced concentrations of 5.33 mg/ml. Restacking is an issue with this technique.

Adding a surfactant to a solvent prior to sonication prevents restacking by adsorbing to the graphene's surface. This produces a higher graphene concentration, but removing the surfactant requires chemical treatments.

Sonicating graphite at the interface of two immiscible liquids, most notably heptane and water, produced macro-scale graphene films. The graphene sheets are adsorbed to the high energy interface between the materials and are kept from restacking. The sheets are up to about 95% transparent and conductive.

With definite cleavage parameters, the box-shaped graphene (BSG) nanostructure can be prepared on graphite crystal.

Molten Salts

Graphite particles can be corroded in molten salts to form a variety of carbon nanostructures including graphene. Hydrogen cations, dissolved in molten lithium chloride, can be discharged on cathodically polarized graphite rods, which then intercalate, peeling graphene sheets. The graphene nanosheets produced displayed a single-crystalline structure with a lateral size of several hundred nanometers and a high degree of crystallinity and thermal stability.

Electrochemical synthesis

Electrochemical synthesis can exfoliate graphene. Varying a pulsed voltage controls thickness, flake area, number of defects and affects its properties. The process begins by bathing the graphite in a solvent for intercalation. The process can be tracked by monitoring the solution's transparency with an LED and photodiode.

Hydrothermal Self-assembly

Graphene has been prepared by using a sugar (e.g. glucose, sugar, fructose, etc.) This substrate-free "bottom-up" synthesis is safer, simpler and more environmentally friendly than exfoliation. The method can control thickness, ranging from monolayer to multilayers, which is known as "Tang-Lau Method".

Chemical Vapor Deposition

Epitaxy

Epitaxial graphene may be coupled to surfaces weakly enough (by Van der Waals forces) to retain the two dimensional electronic band structure of isolated graphene.

Heating silicon carbide (SiC) to high temperatures (1100 °C) under low pressures (c. 10^{-6} torr) reduces it to graphene.

A normal silicon wafer coated with a layer of germanium (Ge) dipped in dilute hydrofluoric acid strips the naturally forming germanium oxide groups, creating hydrogen-terminated germanium. CVD can coat that with graphene.

The direct synthesis of graphene on insulator TiO_2 with high-dielectric-constant (high-κ). A two-step CVD process is shown to grow graphene directly on TiO_2 crystals or exfoliated TiO_2 nanosheets without using any metal catalyst.

Metal Substrates

The atomic structure of metal substrates including ruthenium, iridium, nickel and copper.

Sodium Ethoxide Pyrolysis

Gram-quantities were produced by the reduction of ethanol by sodium metal, followed by pyrolysis and washing with water.

Roll-to-roll

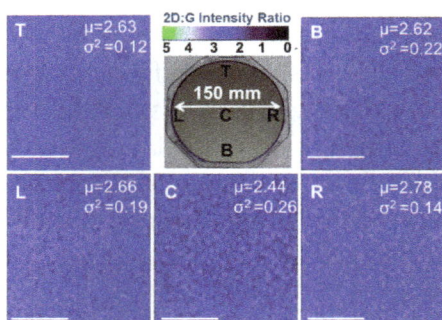

Large-area Raman mapping of CVD graphene on deposited Cu thin film on 150 mm SiO2/Si wafers reveals >95% monolayer continuity and an average value of ~2.62 for I_{2D}/I_G. The scale bar is 200 μm.

In 2014 a two-step roll-to-roll manufacturing process was announced. The first roll-to-roll step produces the graphene via chemical vapor deposition. The second step binds the graphene to a substrate.

Cold Wall

Growing graphene in an industrial resistive-heating cold wall CVD system was claimed to produce graphene 100 times faster than conventional CVD systems, cut costs by 99% and produce material with enhanced electronic qualities.

Wafer Scale CVD Graphene

CVD graphene is scalable and has been integrated with ubiquitous CMOS technology via growth on deposited Cu thin film catalyst on 100 to 300 mm standard Si/SiO2 wafers on an Axitron Black Magic system. Monolayer graphene coverage of >95% is achieved on 100 to 300 mm wafer substrates with negligible defects, confirmed by extensive Raman mapping.

Nanotube Slicing

Graphene can be created by opening carbon nanotubes by cutting or etching. In one such method multi-walled carbon nanotubes are cut open in solution by action of potassium permanganate and sulfuric acid.

Carbon Dioxide Reduction

A highly exothermic reaction combusts magnesium in an oxidation–reduction reaction with carbon dioxide, producing carbon nanoparticles including graphene and fullerenes.

Spin Coating

In 2014, carbon nanotube-reinforced graphene was made via spin coating and annealing functionalized carbon nanotubes.

Supersonic Spray

Supersonic acceleration of droplets through a Laval nozzle was used to deposit reduced graphene-oxide on a substrate. The energy of the impact rearranges that carbon atoms into flawless graphene.

Another approach sprays buckyballs at supersonic speeds onto a substrate. The balls cracked open upon impact, and the resulting unzipped cages then bond together to form a graphene film.

Laser

In 2014 a CO_2 infrared laser produced and patterned porous three-dimensional graphene film networks from commercial polymer films. The result exhibits high electrical conductivity. Laser-induced production appeared to allow roll-to-roll manufacturing processes.

Microwave-assisted Oxidation

In 2012, microwave energy was reported to directly synthesize graphene in one step. This approach avoids use of potassium permanganate in the reaction mixture. It was also reported that by microwave radiation assistance, graphene oxide with or without holes can be synthesized by controlling microwave time. Microwave heating can dramatically shorten the reaction time from days to seconds.

Ion Implantation

Accelerating carbon ions under an electrical field into a semiconductor made of thin nickel films on a substrate of SiO_2/Si, creates a wafer-scale (4 inches (100 mm)) wrinkle/tear/residue-free graphene layer at a relatively low temperature of 500 °C.

Graphene Analogs

Graphene analogs (also referred to as "artificial graphene") are two-dimensional systems which exhibit similar properties to graphene. Graphene analogs are studied intensively since the discovery of graphene in 2004. People try to develop systems in which the physics is easier to observe and to manipulate than in graphene. In those systems, electrons are not always the particles which are used. They might be optical photons, microwave photons, plasmons, microcavity polaritons, or even atoms. Also, the honeycomb structure in which those particles evolve can be of a different nature than carbon atoms in graphene. It can be, respectively, a photonic crystal, an array of metallic rods, metallic nanoparticles, a lattice of coupled microcavities, or an optical lattice.

Applications

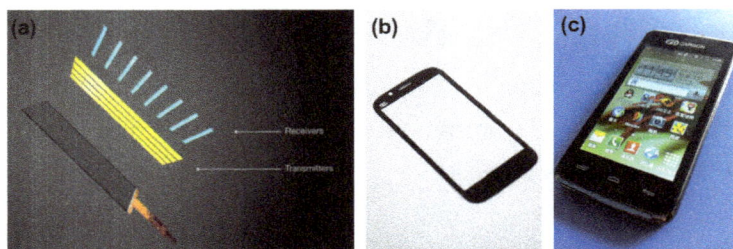

(a) The typical structure of a touch sensor in a touch panel. (b) An actual example of 2D Carbon Graphene Material Co.,Ltd's graphene transparent conductor-based touchscreen that is employed in (c) a commercial smartphone.

Graphene is a transparent and flexible conductor that holds great promise for various material/device applications, including solar cells, light-emitting diodes (LED), touch panels, and smart windows or phones. According to information from Changzhou, China-based 2D Carbon Graphene Material Co.,Ltd, graphene-based touch panel modules have been sold in volume to cell phone, wearable device, and home appliance manufacturers. For instance, smartphone products with graphene touch screens are already on the market.

In 2013, Head announced their new range of graphene tennis racquets.

As of 2015, there is one product available for commercial use: a graphene-infused printer powder. Many other uses for graphene have been proposed or are under development, in areas including electronics, biological engineering, filtration, lightweight/strong composite materials, photovoltaics and energy storage. Graphene is often produced as a powder and as a dispersion in a polymer matrix. This dispersion is supposedly suitable for advanced composites, paints and coatings, lubricants, oils and functional fluids, capacitors and batteries, thermal management applications, display materials and packaging, solar cells, inks and 3D-printers' materials, and barriers and films.

In 2016, researchers have been able to make a graphene film that can absorb 95% of light incident on it. It is also getting cheaper; recently scientists at the University of Glasgow have produced graphene at a cost that is 100 times less than the previous methods.

In August 2, 2016, BAC's new Mono model is said to be made out of graphene as a first of both a street-legal track car and a production car.

Health Risks

The toxicity of graphene has been extensively debated in the literature. The most comprehensive review on graphene toxicity published by Lalwani et al. exclusively summarizes the in vitro, in vivo, antimicrobial and environmental effects and highlights the various mechanisms of graphene toxicity. Results show that the toxicity of graphene is dependent on several factors such as shape, size, purity, post-production processing steps, oxidative state, functional groups, dispersion state, synthesis methods, route and dose of administration, and exposure times.

Research at Stony Brook University showed that graphene nanoribbons, graphene nanoplatelets and graphene nano–onions are non-toxic at concentrations up to 50 μg/ml. These nanoparticles do not alter the differentiation of human bone marrow stem cells towards osteoblasts (bone) or adipocytes (fat) suggesting that at low doses graphene nanoparticles are safe for biomedical applications. Research at Brown university found that 10 μm few-layered graphene flakes are able to pierce cell membranes in solution. They were observed to enter initially via sharp and jagged points, allowing graphene to be internalized in the cell. The physiological effects of this remain uncertain, and this remains a relatively unexplored field.

NANOCOMPOSITE

Nanocomposite is a multiphase solid material where one of the phases has one, two or three dimensions of less than 100 nanometers (nm) or structures having nano-scale repeat distances between the different phases that make up the material.

The idea behind Nanocomposite is to use building blocks with dimensions in nanometre range to design and create new materials with unprecedented flexibility and improvement in their physical properties.

In the broadest sense this definition can include porous media, colloids, gels and copolymers, but is more usually taken to mean the solid combination of a bulk matrix and nano-dimensional phase(s) differing in properties due to dissimilarities in structure and chemistry. The mechanical, electrical, thermal, optical, electrochemical, catalytic properties of the nanocomposite will differ markedly from that of the component materials. Size limits for these effects have been proposed:

1. <5 nm for catalytic activity.

2. <20 nm for making a hard magnetic material soft.

3. <50 nm for refractive index changes.

4. <100 nm for achieving superparamagnetism, mechanical strengthening or restricting matrix dislocation movement.

Nanocomposites are found in nature, for example in the structure of the abalone shell and bone. The use of nanoparticle-rich materials long predates the understanding of the physical and chemical nature of these materials. Jose-Yacaman *et al.* investigated the origin of the depth of colour and the resistance to acids and bio-corrosion of Maya blue paint, attributing it to a nanoparticle mechanism. From the mid-1950s nanoscale organo-clays have been used to control flow of polymer solutions (e.g. as paint viscosifiers) or the constitution of gels (e.g. as a thickening substance in cosmetics, keeping the preparations in homogeneous form).

In mechanical terms, nanocomposites differ from conventional composite materials due to the exceptionally high surface to volume ratio of the reinforcing phase and/or its exceptionally high aspect ratio. The reinforcing material can be made up of particles (e.g. minerals), sheets (e.g. exfoliated clay stacks) or fibres (e.g. carbon nanotubes or electrospun fibres). The area of the interface between the matrix and reinforcement phase(s) is typically an order of magnitude greater than for conventional composite materials. The matrix material properties are significantly affected in the vicinity of the reinforcement. Ajayan *et al.* note that with polymer nanocomposites, properties related to local chemistry, degree of thermoset cure, polymer chain mobility, polymer chain conformation, degree of polymer chain ordering or crystallinity can all vary significantly and continuously from the interface with the reinforcement into the bulk of the matrix.

This large amount of reinforcement surface area means that a relatively small amount of nanoscale reinforcement can have an observable effect on the macroscale properties of the composite. For example, adding carbon nanotubes improves the electrical and thermal conductivity. Other kinds of nanoparticulates may result in enhanced optical properties, dielectric properties, heat resistance or mechanical properties such as stiffness, strength and resistance to wear and damage. In general, the nano reinforcement is dispersed into the matrix during processing. The percentage by weight (called *mass fraction*) of the nanoparticulates introduced can remain very low (on the order of 0.5% to 5%) due to the low filler percolation threshold, especially for the most commonly used non-spherical, high aspect ratio fillers (e.g. nanometer-thin platelets, such as clays, or nanometer-diameter cylinders, such as carbon nanotubes). The orientation and arrangement of asymmetric nanoparticles, thermal property mismatch at the interface, interface density per unit volume of nanocomposite, and polydispersity of nanoparticles significantly affect the effective thermal conductivity of nanocomposites.

Ceramic-matrix Nanocomposites

Ceramic matrix composites (CMCs) consist of ceramic fibers embedded in a ceramic matrix. The matrix and fibers can consist of any ceramic material, including carbon and carbon fibers. The ceramic occupying most of the volume is often from the group of oxides, such as nitrides, borides, silicides, whereas the second component is often a metal. Ideally both components are finely dispersed in each other in order to elicit particular optical, electrical and magnetic properties as well as tribological, corrosion-resistance and other protective properties.

The binary phase diagram of the mixture should be considered in designing ceramic-metal nano-composites and measures have to be taken to avoid a chemical reaction between both components. The last point mainly is of importance for the metallic component that may easily react with the ceramic and thereby lose its metallic character. This is not an easily obeyed constraint because the preparation of the ceramic component generally requires high process temperatures. The safest measure thus is to carefully choose immiscible metal and ceramic phases. A good example of such a combination is represented by the ceramic-metal composite of TiO_2 and Cu, the mixtures of which were found immiscible over large areas in the Gibbs' triangle of Cu-O-Ti.

The concept of ceramic-matrix nanocomposites was also applied to thin films that are solid layers of a few nm to some tens of μm thickness deposited upon an underlying substrate and that play an important role in the functionalization of technical surfaces. Gas flow sputtering by the hollow cathode technique turned out as a rather effective technique for the preparation of nanocomposite layers. The process operates as a vacuum-based deposition technique and is associated with high deposition rates up to some μm/s and the growth of nanoparticles in the gas phase. Nanocomposite layers in the ceramics range of composition were prepared from TiO_2 and Cu by the hollow cathode technique that showed a high mechanical hardness, small coefficients of friction and a high resistance to corrosion.

Metal-matrix Nanocomposites

Metal matrix nanocomposites can also be defined as reinforced metal matrix composites. This type of composites can be classified as continuous and non-continuous reinforced materials. One of the more important nanocomposites is Carbon nanotube metal matrix composites, which is an emerging new material that is being developed to take advantage of the high tensile strength and electrical conductivity of carbon nanotube materials. Critical to the realization of CNT-MMC possessing optimal properties in these areas are the development of synthetic techniques that are (a) economically producible, (b) provide for a homogeneous dispersion of nanotubes in the metallic matrix, and (c) lead to strong interfacial adhesion between the metallic matrix and the carbon nanotubes. In addition to carbon nanotube metal matrix composites, boron nitride reinforced metal matrix composites and carbon nitride metal matrix composites are the new research areas on metal matrix nanocomposites.

A recent study, comparing the mechanical properties (Young's modulus, compressive yield strength, flexural modulus and flexural yield strength) of single- and multi-walled reinforced polymeric (polypropylene fumarate—PPF) nanocomposites to tungsten disulfide nanotubes reinforced PPF nanocomposites suggest that tungsten disulfide nanotubes reinforced PPF nanocomposites possess significantly higher mechanical properties and tungsten disulfide nanotubes are better reinforcing agents than carbon nanotubes. Increases in the mechanical properties can be attributed to a uniform dispersion of inorganic nanotubes in the polymer matrix (compared to carbon nanotubes that exist as micron sized aggregates) and increased crosslinking density of the polymer in the presence of tungsten disulfide nanotubes (increase in crosslinking density leads to an increase in the mechanical properties). These results suggest that inorganic nanomaterials, in general, may be better reinforcing agents compared to carbon nanotubes.

Another kind of nanocomposite is the energetic nanocomposite, generally as a hybrid sol–gel with a silica base, which, when combined with metal oxides and nano-scale aluminum powder, can form *superthermite* materials.

Polymer-matrix Nanocomposites

In the simplest case, appropriately adding nanoparticulates to a polymer matrix can enhance its performance, often dramatically, by simply capitalizing on the nature and properties of the nanoscale filler (these materials are better described by the term *nanofilled polymer composites*). This strategy is particularly effective in yielding high performance composites, when uniform dispersion of the filler is achieved and the properties of the nanoscale filler are substantially different or better than those of the matrix. The uniformity of the dispersion is in all nanocomposites is counteracted by thermodynamically driven phase separation. Clustering of nanoscale fillers produces aggregates that serve as structural defects and result in failure. Layer-by-layer (LbL) assembly when nanometer scale layers of nanoparticulates and a polymers are added one by one. LbL composites display performance parameters 10-1000 times better that the traditional nanocomposites made by extrusion or batch-mixing.

Nanoparticles such as graphene, carbon nanotubes, molybdenum disulfide and tungsten disulfide are being used as reinforcing agents to fabricate mechanically strong biodegradable polymeric nanocomposites for bone tissue engineering applications. The addition of these nanoparticles in the polymer matrix at low concentrations (~0.2 weight %) cause significant improvements in the compressive and flexural mechanical properties of polymeric nanocomposites. Potentially, these nanocomposites may be used as a novel, mechanically strong, light weight composite as bone implants. The results suggest that mechanical reinforcement is dependent on the nanostructure morphology, defects, dispersion of nanomaterials in the polymer matrix, and the cross-linking density of the polymer. In general, two-dimensional nanostructures can reinforce the polymer better than one-dimensional nanostructures, and inorganic nanomaterials are better reinforcing agents than carbon based nanomaterials. In addition to mechanical properties, polymer nanocomposites based on carbon nanotubes or graphene have been used to enhance a wide range of properties, giving rise to functional materials for a wide range of high added value applications in fields such as energy conversion and storage, sensing and biomedical tissue engineering. For example, multi-walled carbon nanotubes based polymer nanocomposites have been used for the enhancement of the electrical conductivity.

Nanoscale dispersion of filler or controlled nanostructures in the composite can introduce new physical properties and novel behaviors that are absent in the unfilled matrices. This effectively changes the nature of the original matrix (such composite materials can be better described by the term *genuine nanocomposites* or *hybrids*). Some examples of such new properties are fire resistance or flame retardancy, and accelerated biodegradability.

A range of polymeric nanocomposites are used for biomedical applications such as tissue engineering, drug delivery, cellular therapies. Due to unique interactions between polymer and nanoparticles, a range of property combinations can be engineered to mimic native tissue structure and properties. A range of natural and synthetic polymers are used to design polymeric nanocomposites for biomedical applications including starch, cellulose, alginate, chitosan, collagen, gelatin, and fibrin, poly(vinyl alcohol) (PVA), poly(ethylene glycol) (PEG), poly(caprolactone) (PCL), poly(lactic-co-glycolic acid) (PLGA), and poly(glycerol sebacate) (PGS). A range of nanoparticles including ceramic, polymeric, metal oxide and carbon-based nanomaterials are incorporated within polymeric network to obtain desired property combinations.

Magnetic Nanocomposites

Nanocomposites that can respond to an external stimulus are of increased interest due to the fact that, due to the large amount of interaction between the phase interfaces, the stimulus response can have a larger effect on the composite as a whole. The external stimulus can take many forms, such as a magnetic, electrical, or mechanical field. Specifically, magnetic nanocomposites are useful for use in these applications due to the nature of magnetic material's ability to respond both to electrical and magnetic stimuli. The penetration depth of a magnetic field is also high, leading to an increased area that the nanocomposite is affected by and therefore an increased response. In order to respond to a magnetic field, a matrix can be easily loaded with nanoparticles or nanorods The different morphologies for magnetic nanocomposite materials are vast, including matrix dispersed nanoparticles, core-shell nanoparticles, colloidal crystals, macroscale spheres, or janus-type nanostructures.

Magnetic nanocomposites can be utilized in a vast number of applications, including catalytic, medical, and technical. For example, palladium is a common transition metal used in catalysis reactions. Magnetic nanoparticle-supported palladium complexes can be used in catalysis to increase the efficiency of the palladium in the reaction.

Magnetic nanocomposites can also be utilized in the medical field, with magnetic nanorods embedded in a polymer matrix can aid in more precise drug delivery and release. Finally, magnetic nanocomposites can be used in high frequency/high temperature applications. For example, multi-layer structures can be fabricated for use in electronic applications. An electrodeposited Fe/Fe oxide multi-layered sample can be an example of this application of magnetic nanocomposites.

NANOSTRUCTURED FILM

A nanostructured film is a film resulting from engineering of nanoscale features, such as dislocations, grain boundaries, defects, or twinning. In contrast to other nanostructures, such as nanoparticles, the film itself may be up to several microns thick, but possesses a large concentration of nanoscale features homogeneously distributed throughout the film. Like other nanomaterials, nanostructured films have sparked much interest as they possess unique properties not found in bulk, non-nanostructured material of the same composition. In particular, nanostructured films have been the subject of recent research due to their superior mechanical properties, including strength, hardness, and corrosion resistance compared to regular films of the same material. Examples of nanostructured films include those produced by grain boundary engineering, such as nano-twinned ultra-fine grain copper, or dual phase nanostructuring, such as crystalline metal and amorphous metallic glass nanocomposites.

Synthesis and Characterization

Nanostructured films are commonly created using magnetron sputtering from an appropriate target material. Films can be elemental in nature, formed by sputtering from a pure metal target such as copper, or composed of compound materials. Varying parameters such as the sputtering rate, substrate temperature, and sputtering interrupts allow the creation of films with a variety of different nanostructured elements. Control over nano-twinning, tailoring of specific types of grain boundaries, and restricting the movement and propagation of dislocations have been demonstrated using films produced via magnetron sputtering.

Methods used to characterize nanostructured films include transmission electron microscopy, scanning electron microscopy, electron backscatter diffraction, focused ion beam milling, and nanoindentation. These techniques are used as they allow imaging of nanoscale structures, including dislocations, twinning, grain boundaries, film morphology, and atomic structure.

Material Properties

Nanostructured films are of interest due to their superior mechanical and physical properties compared to their normal equivalent. Elemental nanostructured films composed of pure copper were found to possess good thermal stability due to the nano-twinned film possessing a higher fraction of grain boundaries. In addition to possessing higher thermal stability, copper films that were highly nano-twinned were found to have a better corrosion resistance than copper films with a low concentration of nano-twins. Control of the fraction of grains in a material with nano-twins present has great potential for less expensive alloys and coatings with a good degree of corrosion resistance.

Compound nanostructured films composed of crystalline $MgCu_2$ cores encapsulated by amorphous glassy shells of the same material were shown to possess a near-ideal mechanical strength. The crystalline $MgCu_2$ cores, typically less than 10 nm in size, were found to substantially strengthen the material by restricting the movement of dislocations and grains. The cores were also found to contribute to overall material strength by restricting the movement of shear bands in the material. This nanostructured film differs from both crystalline metals and amorphous metallic glasses, both of which exhibit behaviors such as the reverse Hall-Petch and shear-band softening effects that prevent them from reaching ideal strength values.

Applications

Nanostructured films with superior mechanical properties allow previously unusable materials to be utilized in new applications, enabling advances fields where coatings are heavily utilized, such as aerospace, energy, and other engineering fields. Production scalability of nanostructured films has already been demonstrated, and the ubiquity of sputtering techniques in industry is predicted to facilitate the incorporation of nanostructured films into existing applications.

APPLICATIONS OF NANOMATERIALS

Since nanomaterials possess unique, beneficial chemical, physical, and mechanical properties, they can be used for a wide variety of applications. These applications include, but are not limited to, the following:

Next-generation Computer Chips

The microelectronics industry has been emphasising miniaturisation, whereby the circuits, such as transistors, resistors, and capacitors, are reduced in size. By achieving a significant reduction in their size, the microprocessors, which contain these components, can run much faster, thereby enabling computations at far greater speeds. However, there are several technological impediments to these advancements, including lack of the ultrafine precursors to manufacture these components; poor dissipation of tremendous amount of heat generated by these microprocessors due to faster speeds; short mean time to failures (poor reliability), etc. Nanomaterials help the industry break these barriers down by providing the manufacturers with nanocrystalline starting materials, ultra-high purity materials, materials with better thermal conductivity, and longer-lasting, durable interconnections (connections between various components in the microprocessors).

Better Insulation Materials

Nanocrystalline materials synthesised by the sol-gel technique result in foam like structures called "aerogels." These aerogels are porous and extremely lightweight; yet, they can loads equivalent to 100 times their weight. Aerogels are composed of three-dimensional, continuous networks of particles with air (or any other fluid, such as a gas) trapped at their interstices. Since they are porous and air is trapped at the interstices, aerogels are currently being used for insulation in offices, homes, etc. By using aerogels for insulation, heating and cooling bills are drastically reduced, thereby saving power and reducing the attendant environmental pollution. They are also being used as materials for "smart " windows, which darken when the sun is too bright (just as in changeable lenses in prescription spectacles and sunglasses) and they lighten themselves, when the sun is not shining too brightly.

Phosphors for High-definition TV

The resolution of a television, or a monitor, depends greatly on the size of the pixel. These pixels are essentially made of materials called "phosphors," which glow when struck by a stream of electrons inside the cathode ray tube (CRT). The resolution improves with a reduction in the size of the pixel, or the phosphors. Nanocrystalline zinc selenide, zinc sulfide, cadmium sulfide, and lead telluride synthesised by the sol-gel techniques are candidates for improving the resolution of monitors. The use of nanophosphors is envisioned to reduce the cost of these displays so as to render high-definition televisions (HDTVs) and personal computers affordable to be purchased by an average household in the U. S.

Low-cost Flat-panel Displays

Flat-panel displays represent a huge market in the laptop (portable) computers industry. However, Japan is leading this market, primarily because of its research and development efforts on the materials for such displays. By synthesising nanocrystalline phosphors, the resolution of these display devices can be greatly enhanced, and the manufacturing costs can be significantly reduced. Also, the flat-panel displays constructed out of nanomaterials possess much higher brightness and contrast than the conventional ones owing to their enhanced electrical and magnetic properties.

Tougher and Harder Cutting Tools

Cutting tools made of nanocrystalline materials, such as tungsten carbide, tantalum carbide, and titanium carbide, are much harder, much more wear-resistant, erosion-resistant, and last longer than their conventional (large-grained) counterparts. They also enable the manufacturer to machine various materials much faster, thereby increasing productivity and significantly reducing manufacturing costs. Also, for the miniaturisation of microelectronic circuits, the industry needs microdrills (drill bits with diameter less than the thickness of an average human hair or 100 μm) with enhanced edge retention and far better wear resistance. Since nanocrystalline carbides are much stronger, harder, and wear-resistant, they are currently being used in these microdrills.

Elimination of Pollutants

Nanocrystalline materials possess extremely large grain boundaries relative to their grain size. Hence, nanomaterials are very active in terms of their of chemical, physical, and mechanical properties. Due to their enhanced chemical activity, nanomaterials can be used as catalysts to react with such noxious and toxic gases as carbon monoxide and nitrogen oxide in automobile catalytic converters and power generation equipment to prevent environmental pollution arising from burning gasoline and coal.

High Energy Density Batteries

Conventional and rechargeable batteries are used in almost all applications that require electric power. These applications include automobiles, laptop computers, electric vehicles, next-generation electric vehicles (NGEV) to reduce environmental pollution, personal stereos, cellular phones, cordless phones, toys, and watches. The energy density (storage capacity) of these batteries is quite low requiring frequent recharging. The life of conventional and rechargeable batteries is also low. Nanocrystalline materials synthesised by sol-gel techniques are candidates for separator plates in batteries because of their foam-like (aerogel) structure, which can hold considerably more energy than their conventional counterparts. Furthermore, nickel-metal hydride (Ni-MH) batteries made of nanocrystalline nickel and metal hydrides are envisioned to require far less frequent recharging and to last much longer because of their large grain boundary (surface) area and enhanced physical, chemical, and mechanical properties.

High-power Magnets

The strength of a magnet is measured in terms of coercivity and saturation magnetisation values. These values increase with a decrease in the grain size and an increase in the specific surface area (surface area per unit volume of the grains) of the grains. It has been shown that magnets made of nanocrystalline yttrium-samarium-cobalt grains possess very unusual magnetic properties due to their extremely large surface area. Typical applications for these high-power rare-earth magnets include quieter submarines, automobile alternators, land-based power generators, motors for ships, ultra-sensitive analytical instruments, and magnetic resonance imaging (MRI) in medical diagnostics.

High-sensitivity Sensors

Sensors employ their sensitivity to the changes in various parameters they are designed to measure. The measured parameters include electrical resistivity, chemical activity, magnetic permeability, thermal conductivity, and capacitance. All of these parameters depend greatly on the microstructure (grain size) of the materials employed in the sensors. A change in the sensor's environment is manifested by the sensor material's chemical, physical, or mechanical characteristics, which is exploited for detection. For instance, a carbon monoxide sensor made of zirconium oxide (zirconia) uses its chemical stability to detect the presence of carbon monoxide. In the event of carbon monoxide's presence, the oxygen atoms in zirconium oxide react with the carbon in carbon monoxide to partially reduce zirconium oxide. This reaction triggers a change in the sensor's characteristics, such as conductivity (or resistivity) and capacitance. The rate and the extent of this reaction are greatly increased by a decrease in the grain size. Hence, sensors made nanocrystalline materials are extremely sensitive to the change in their environment. Typical applications for sensors made out of nanocrystalline materials are smoke detectors, ice detectors on aircraft wings, automobile engine performance sensor, etc.

Automobiles with Greater Fuel Efficiency

Currently, automobile engines waste considerable amounts of gasoline, thereby contributing to environmental pollution by not completely combusting the fuel. A conventional spark plug is not designed to burn the gasoline completely and efficiently. This problem is compounded by defective, or worn-out, spark plug electrodes. Since nanomaterials are stronger, harder, and much more wear-resistant and erosion-resistant, they are presently being envisioned to be used as spark plugs. These electrodes render the spark plugs longer-lasting and combust fuel far more efficiently and completely. A radically new spark plug design called the "railplug" is also in the prototype stages. This railplug uses the technology derived from the "railgun," which is a spin-off of the popular Star Wars defense program. However, these railplugs generate much more powerful sparks (with an energy density of approximately 1 kJ/mm^2). Hence, conventional materials erode and corrode too soon and quite frequently to be of any practical use in automobiles. Nevertheless, railplugs made of nanomaterials last much longer even the conventional spark plugs. Also, automobiles waste significant amounts of energy by losing the thermal energy generated by the engine. This is especially true in the case of diesel engines. Hence, the engine cylinders (liners) are currently being envisioned to be coated with nanocrystalline ceramics, such as zirconia and alumina, so that they retain heat much more efficiently and result in complete and efficient combustion of the fuel.

Aerospace Components with Enhanced Performance Characteristics

Due to the risks involved in flying, aircraft manufacturers strive to make the aerospace components stronger, tougher, and last longer. One of the key properties required of the aircraft components is the fatigue strength, which decreases with the component's age. By making the components out of stronger materials, the life of the aircraft is greatly increased. The fatigue strength increases with a reduction in the grain size of the material. Nanomaterials provide such a significant reduction in the grain size over conventional materials that the fatigue life is increased by an average of 200-300%. Furthermore, components made of nanomaterials are stronger and can operate at higher temperatures, aircrafts can fly faster and more efficiently (for the same amount of aviation fuel). In spacecrafts, elevated-temperature strength of the material is crucial because the components (such as rocket engines, thrusters, and vectoring nozzles) operate at much higher temperatures than aircrafts and higher speeds. Nanomaterials are perfect candidates for spacecraft applications, as well.

Better and Future Weapons Platforms

Conventional guns, such as cannons, 155 mm howitzers, and multiple-launch rocket system

(MLRS), utilise the chemical energy derived by igniting a charge of chemicals (gun powder). The maximum velocity at which the penetrator can be propelled is approximately 1.5-2.0 km/sec. On the other hand, electromagnetic launchers (EML guns), or railguns, use the electrical energy, and the concomitant magnetic field (energy), to propel the penetrators/projectiles at velocities up to 10 km/sec. This increase in velocity results in greater kinetic energy for the same penetrator mass. The greater the energy, the greater is the damage inflicted on the target. For this and other reasons, the DoD (especially, the U. S. Army) has conducted extensive research into the railguns. Since a railgun operates on electrical energy, the rails need to be very good conductors of electricity. Also, they need to be so strong and rigid that the railgun does not sag while firing and buckle under its own weight. The obvious choice for high electrical conductivity is copper. However, the railguns made out of copper wear out much too quickly due to the erosion of the rails by the hypervelocity projectiles and they lack high-temperature strength. The wear and erosion of copper rails necessitate inordinately frequent barrel replacements. In order to satisfy these requirements, a nanocrystalline composite material made of tungsten, copper, and titanium diboride is being evaluated as a potential candidate. This nanocomposite possesses the requisite electrical conductivity, adequate thermal conductivity, excellent high strength, high rigidity, hardness, and wear/erosion resistance. This results in longer-lasting, wear-resistant, and erosion-resistant railguns, which can be fired more frequently and often than their conventional counterparts.

Longer-lasting Satellites

Satellites are being used for both defence and civilian applications. These satellites utilise thruster rockets to remain in or change their orbits due to a variety of factors including the influence of gravitational forces exerted by the earth. Hence, these satellites are repositioned using these thrusters. The life of these satellites, to a large extent, is determined by the amount of fuel they can carry on board. In fact, more than 1/3 of the fuel carried aboard by the satellites is wasted by these repositioning thrusters due to incomplete and inefficient combustion of the fuel, such as hydrazine. The reason for the incomplete and inefficient combustion is that the onboard ignitors wear out quickly and cease to perform effectively. Nanomaterials, such as nanocrsytalline tungsten-titanium diboride-copper composite, are potential candidates for enhancing these ignitors' life and performance characteristics.

Longer-lasting Medical Implants

Currently, medical implants, such as orthopaedic implants and heart valves, are made of titanium and stainless steel alloys. These alloys are primarily used in humans because they are biocompatible, i.e., they do not adversely react with human tissue. In the case of orthopaedic implants (artificial bones for hip, etc.), these materials are relatively non-porous. For an implant to effectively mimic a natural human bone, the surrounding tissue must penetrate the implants, thereby affording the implant with the required strength. Since these materials are relatively impervious, human tissue does not penetrate the implants, thereby reducing their effectiveness. Furthermore, these metal alloys wear out quickly necessitating frequent, and often very expensive, surgeries. However, nanocrystalline zirconia (zirconium oxide) ceramic is hard, wear-resistant, corrosion-resistant (biological fluids are corrosive), and biocompatible. Nanoceramics can also be made porous into aerogels (aerogels can withstand up to 100 times their weight), if they are synthesized by sol-gel techniques. This results in far less frequent implant replacements, and hence, a significant reduction in surgical expenses. Nanocrystalline silicon carbide (SiC) is a candidate material for artificial heart valves primarily due to its low weight, high strength, extreme hardness, wear resistance, inertness (SiC does not react with biological fluids), and corrosion resistance.

Ductile and Machinable Ceramics

Ceramics, per se, are very hard, brittle, and hard to machine. These characteristics of ceramics have discouraged the potential users from exploiting their beneficial properties. However, with a reduction in grain size, these ceramics have increasingly been used. Zirconia, a hard, brittle ceramic, has even been rendered superplastic, i.e., it can be deformed to great lengths (up to 300% of its original length). However, these ceramics must possess nanocrystalline grains to be superplastic. In fact, nanocrystalline ceramics, such as silicon nitride (Si_3N_4) and silicon carbide (SiC), have been used in such automotive applications as high-strength springs, ball bearings, and valve lifters, because they possess good formability and machinabilty combined with excellent physical, chemical, and mechanical properties. They are also used as components in high-temperature furnaces. Nanocrystalline ceramics can be pressed and sintered into various shapes at significantly lower temperatures, whereas it would be very difficult, if not impossible, to press and sinter conventional ceramics even at high temperatures.

Large Electrochromic Display Devices

An electrochromic device consists of materials in which an optical absorption band can be introduced, or an existing band can be altered by the passage of current through the materials, or by the application of an electric field. Nanocrystalline materials, such as tungstic oxide ($WO_3.xH_2O$) gel, are used in very large electrochromic display devices. The reaction governing electrochromism (a reversible coloration process under the influence of an electric field) is the double-injection of ions (or protons, H^+) and electrons, which combine with the nanocrystalline tungstic acid to form a tungsten bronze. These devices are primarily used in public billboards and ticker boards to convey information. Electrochromic devices are similar to liquid-crystal displays (LCD) commonly used in calculators and watches. However, electrochromic devices display information by changing colour when a voltage is applied. When the polarity is reversed, the colour is bleached. The resolution, brightness, and contrast of these devices greatly depend on the tungstic acid gel's grain size. Hence, nanomaterials are being explored for this purpose.

References

- E., Shinn; A., Hubler; D., Lyon; M., Grosse-Perdekamp; A., Bezryadin; A., Belkin (22 October 2012). "Nuclear Energy Conversion with Stacks of Graphene Nano-capacitors". Complexity. 18 (3): 24–27. Doi:10.1002/cplx.21427

- Nanoparticle, science: britannica.com, Retrieved 10 February, 2019

- "Global Demand for Graphene after Commercial Production to be Enormous, says Report". AZONANO.com. 28 February 2014. Retrieved 24 July 2014

- "Graphene layer". IUPAC Compendium of Chemical Terminology. International Union of Pure and Applied Chemistry. 2009. doi:10.1351/goldbook.G02683. ISBN 978-0-9678550-9-7. Retrieved 31 March 2012

- Polyakov, Mikhail N.; Chookajorn, Tongjai; Mecklenburg, Matthew; Schuh, Christopher A.; Hodge, Andrea M. (2016-04-15). "Sputtered Hf–Ti nanostructures: A segregation and high-temperature stability study". Acta Materialia. 108: 8–16. doi:10.1016/j.actamat.2016.01.073

Permissions

Index

www.ingramcontent.com/pod-product-compliance
Lightning Source LLC
Chambersburg PA
CBHW061255190326
41458CB00011B/3674